SOCIETY FOR EXPERIMENTAL BIOLOGY
seminar series: 30

ROOT DEVELOPMENT AND FUNCTION

Root development and function

Edited by
P. J. GREGORY
Department of Soil Science
University of Reading
J. V. LAKE
Agriculture and Food Research Council Headquarters
London
and
D. A. ROSE
Department of Soil Science
University of Newcastle upon Tyne

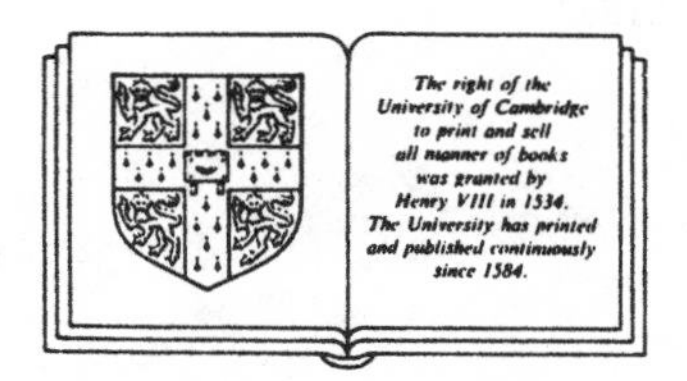

CAMBRIDGE UNIVERSITY PRESS
Cambridge
New York Port Chester
Melbourne Sydney

CAMBRIDGE UNIVERSITY PRESS
Cambridge, New York, Melbourne, Madrid, Cape Town, Singapore, São Paulo, Delhi

Cambridge University Press
The Edinburgh Building, Cambridge CB2 8RU, UK

Published in the United States of America by Cambridge University Press, New York

www.cambridge.org
Information on this title: www.cambridge.org/9780521103640

First published 1987
Reprinted 1988, 1990
This digitally printed version 2009

A catalogue record for this publication is available from the British Library

Library of Congress Cataloguing in Publication data

Root development and function.
(Seminar series; 30)
Based on the review papers presented at the Annual
Meeting of the Environmental Physiology Group of the
Society for Experimental Biology and held jointly with
the Association of Applied Biologists and the British
Ecological Society, at the University College of North
Wales, Bangor, from the 26th to the 28th March, 1985–
Pref.
1. Roots (Botany) – Congresses. I. Gregory, P.J.
II. Lake, J.V. III. Rose, D.A. IV. Series: Seminar
series (Society for Experimental Biology (Great
Britain)); 30.
QK644.R65 1986 581.1′0428 86–21568

ISBN 978-0-521-32931-6 hardback
ISBN 978-0-521-10364-0 paperback

CONTENTS

CONTRIBUTORS

P.W.BARLOW
A.F.R.C. Letcombe Laboratory, Wantage, Oxon, OX12 9JT, U.K.
Present address: Long Ashton Research Station, Department of Agricultural Sciences, University of Bristol, Long Ashton, Bristol BS18 9AF, U.K.

M.M.CALDWELL
Department of Range Science, College of Natural Resources, UMC 52, Utah State University, Logan, Utah, 84322, U.S.A.

M.C.DREW
A.F.R.C. Letcombe Laboratory, Wantage, Oxon, OX12, 9JT, U.K.
Present address: Long Ashton Research Station, Department of Agricultural Sciences, University of Bristol, Long Ashton, Bristol BS18 9AF, U.K.

P.J.GREGORY
Department of Soil Science, University of Reading, London Road, Reading, RG1 5AQ, U.K.

BETTY KLEPPER
U.S.D.A. - A.R.S. Columbia Plateau Conservation Research Center, P.O.Box 370, Pendleton, Oregon, 97801, U.S.A.

P.J.C.KUIPER
Department of Plant Physiology, Biological Centre, University of Groningen, P.O.Box 14, 9750AA Haren, The Netherlands.

J.V.LAKE
Agriculture and Food Research Council, 160 Great Portland Street, London W1N 6DT, U.K.

H.LAMBERS
Department of Plant Ecology, University of Utrecht, Lange Nieuwstraat 106, 3512PN Utrecht, The Netherlands.

WILLIAM J. LUCAS
Forstbotanisches Institut der Universitat Göttingen, 3400 Göttingen, West Germany. Present address: Department of Botany, University of California, Davis, California, 95616, U.S.A.

M.E.McCULLY
Department of Biology, Carleton University, Ottawa, K1S 5B6, Canada.

PREFACE

In planning this volume, we have tried to give equal weight to root development and root function and to treat both in relation to the whole range of physical environmental factors. These clearly include temperature, gravity, light and the mechanical impedance of the soil. But the boundary with the chemical environment is not sharp and our list thus includes water, which interacts with the ionic strength of the major nutrients in the soil solution. We include also soil aeration, which bears on both the access of oxygen and the escape of gases generated in the roots or soil. All these factors vary with depth in the soil and with time, adding greatly to the complexity of the subject.

The physical environment affects also the organisms associated with roots, but as will be seen, experimental evidence on the size of any such effects on mycorrhizae or nitrogen-fixing bacteria is scant.

Where possible, the approach is quantitative, including comparisons between species or between cultivars within a species, partly as a safeguard against generalizing about the probable underlying mechanisms and partly to inform those engaged in plant breeding or genetic manipulation.

After a prologue, on interactions between factors in the physical environment, we consider the effects of these factors on root development and root function, each at four levels of organization - cells, tissues, single plants and plant communities. A concluding synthesis identifies opportunities for closing important gaps in our knowledge of the subject.

This volume is based on the review papers presented at the Annual Meeting of the Environmental Physiology Group of the Society for Experimental Biology and held jointly with the Association of Applied Biologists and the British Ecological Society, at the University College of North Wales, Bangor, from the 26th to the 28th March, 1985. We thank the Group and the two societies for their generous financial support.

We are grateful to the contributors for their carefully prepared manuscripts and their forebearance with our criticism, to Dr. L.D. Incoll (University of Leeds) and Dr. C. Marshall (University College of North Wales) for assistance with organizing the meeting and to staff of the Cambridge University Press.

P.J. Gregory,
J.V. Lake,
D.A. Rose

PROLOGUE: INTERACTIONS IN THE PHYSICAL ENVIRONMENT OF PLANT ROOTS

J.V. Lake

INTRODUCTION

Plant roots modify and interact with each of the physical factors in their environment, but only with gravity is the interaction always negligibly small. Usually roots grow in soil, but occasionally, in experiments or in a natural environment, they will be in air or water. I shall treat each physical factor in turn and indicate its interactions with others and with the plant roots. Many of the factors affect also the growth and function of the plant shoot and this in turn can have second-order effects on the soil, but I shall not attempt to specify them all.

TEMPERATURE

Close to the soil surface, temperature gradients are steep and the amplitude of the diurnal variation is greatest. A gradient of >6 K cm^{-1} in the 2 to 4 cm soil layer and an amplitude of 56 K at 0.4 cm depth were measured by Sinclair in 1922 in the Arizona desert (Sutton, 1953) and have been exceeded in various reports since, although not often convincingly. By contrast, at a depth of 50 cm, i.e. within the rooting range of many plant species, the temperature remains essentially constant by day, although still varying with season. Thus steep and rapidly changing positive or negative temperature gradients can exist along the length of a root, no doubt affecting both structure and function, but most experiments have avoided these complications.

By virtue of their water content, roots will be relatively good conductors of heat compared with the surrounding soil if it is dry, but their heat capacity will be relatively small, so root and soil temperatures are unlikely to differ greatly. A larger interaction of roots with their thermal environment results from the effects of root activity in changing the structure and water content of the surrounding soil.

In the chapters that follow, distinctions will be noted between the effects of temperature on development, growth and function. For example, the shape of the temperature response curve for the cell cycle, or for the induction of lateral primordia, may be quite different from that for cell expansion or root extension. Some of these measured effects (e.g. in relation to so-called thermal time) are reproducible and predictable, but have not yet been subject to biochemical interpretation.

GRAVITY

In our context, the strength and local direction of the gravitational field are almost uniform over the earth's surface. Gravity is only one amongst many factors determining the direction of root growth, but in an otherwise isotropically uniform environment the angle of a particular root to the vertical will depend on its order in the system of branching and on the strength of the gravitropic response of the species or cultivar. Research on the gravitropic response has been especially rewarding in providing increased understanding of growth regulation at the cellular level in roots, although the fundamental nature of the cell-signalling process that elicits the response itself remains unknown.

Environmental factors are rarely independent in their effects on root growth or function. In some species, the strength of the geotropic response depends on the duration, flux density and spectral composition of any light reaching the root apex. The response is minimal in darkness, but brief illumination to allow visual examination of the roots may suffice to elicit gravitropic curvature (Lake & Slack, 1961). The details of this interaction and its spectral dependence are obscure.

LIGHT

In a sandy soil, light may be transmitted through the mineral particles, usually with change in its spectral composition (Mandoli *et al*, 1982), and in all soils light may be transmitted through a film of water or through cracks. The roots themselves may act as light pipes (Mandoli & Briggs, 1984), altering in the process the spectral composition of the light so transmitted, and thus again interacting with their soil environment.

In some plant species, light has no effect on root growth. In others, light of appropriate spectral composition reaching the root apex by whatever means will slow root extension and elicit a gravitropic response (Lake & Slack, *loc. cit.*). Clearly, a competitive advantage may be conferred by the resulting ability of the root system to respond to environmental stimuli other than gravity in the soil and so to extend into zones favourable for growth and function and yet avoid surface layers where the environment may be favourable only transiently.

Light affects root growth indirectly also through regulating plant dry matter production and its partitioning between shoot and root.

MECHANICAL IMPEDANCE

A root extending through soil may force the particles aside, overcoming the mechanical impedance, or may be diverted to travel around the particles if the impedance is too great and if space allows. Similarly, a root may displace or be diverted around an aggregate of small particles. The mechanical impedance will vary between horizons in a soil profile and sometimes within a horizon if the passage of tillage implements or wheels has caused compaction. It will vary also depending on root diameter.

The passage of roots through the soil can alter the subsequent mechanical impedance, first through the effect of root exudates in providing lubrication for the passage of the root tip, second through the subsequent drying by absorption of water for transpiration and third through the creation of channels that can persist and so alter the effective mechanical impedance encountered by succeeding roots.

The conditions in the soil, e.g. dryness and compaction, that tend to increase mechanical impedance also affect access to the roots of air, water and mineral nutrients, so that carefully controlled experiments are necessary to estimate the actual effects of impedance. Although it is then found to slow extension, the evidence for any effect on development at the level of cell differentiation seems equivocal (Barlow, Chapter 1). Yet some effect might be expected because a principal factor contributing to positional recognition for a differentiating cell is thought to be the pattern of mechanical constraint by neighbouring cells (Barlow, 1984).

WATER

Apart from gravity, all the physical factors in the soil are modified by water. Hydraulic conductivity and diffusivity markedly affect the rate of absorption of water and both are complex functions of water content. Different components of the soil water potential near the root surface affect different processes in root growth and function. Steep gradients of both content and potential are common in soils and are a prime cause of gradients in both physical and chemical environmental factors along the length of the root.

Roots interact with soil water by exudation and absorption, but also by shrinking or swelling and so varying the area of their contact with the soil. This process may account for otherwise puzzling experimental observations on root function. Since water is transported to roots down a gradient of soil water potential, instrumentation that would enable potential gradients to be measured close to roots without significantly altering the flow or affecting the environment would advance the pace of research on plant water relations.

As soil dries, the ionic strength of nutrients in the soil solution may increase to a point where the osmotic potential in the soil becomes a significant component of the total water potential, although affecting water uptake only to the extent that the relevant membranes in the root tissue are semi-permeable to the ions concerned. On the other hand, the transport of mineral nutrients to roots in the soil solution may become impaired in drying soil, through a decrease in the diffusion coefficient, causing a deficiency close to the absorbing roots.

SOIL AERATION

The resistance to gas exchange between roots and the atmosphere depends not only on the volume of air-filled pores in the intervening soil layer, but also on the geometry of the pore space. Clearly, both depend on structure and water content. Pore geometry depends also on the ability of the soil to develop fissures through the effects of wetting, drying or frost and on the persistence of these fissures and of channels left (e.g. by the passage of earthworms or plant roots).

Clearly even in compact and flooded soil the partial pressure of oxygen and other gases will be in equilibrium with the atmosphere if the soil is sterile. Gradients of oxygen partial pressure are caused by the respiratory activity of roots and soil organisms; aeration is then a prime example of interaction between roots and soil.

Although the soil itself can be characterised by its diffusive resistance to oxygen transport, the oxygen partial pressure at the root surface depends also on the rate of root respiration; this is why waterlogging in winter may not be disastrous. Under conditions where root respiration is oxygen-limited the oxygen partial pressure at the root surface is presumably close to zero and the oxygen flux density to a sensor at zero potential can be used as a measure of aeration (Blackwell & Wells, 1983).

with sparse rooting, the oxygen supply may originate from zones in the soil unoccupied by respiring organisms, rather than by a direct pathway from the air above the soil.

A decrease in oxygen partial pressure may increase the production of ethylene by soil micro-organisms and plant roots and the diffusion of ethylene away from the production sites depends on the air-filled pore volume and geometry. An increase in the ethylene concentration in plant roots caused by poor aeration can in turn result in the formation of aerenchyma - gas filled spaces within the cortex that facilitate acropetal diffusion of oxygen within the tissue; aeration is then, to some extent, restored.

In some species, poor aeration moderates or reverses root gravitropism, probably through an effect of ethylene. Such a response may be beneficial in bringing roots closer to the usually better-aerated surface soil (Jackson & Drew, 1984).

CONCLUSION

The above examples illustrate that interactions, which are not additive, characterize the relationships of soil physical factors with one another and with plant roots. These interactions should be kept in mind in the following chapters, where attention is turned to the plants themselves at various levels of organization and where inevitably the environmental factors tend to be treated one by one.

REFERENCES

Barlow, P.W. (1984). Positional controls in root development. In *Positional Controls in Plant Development*, ed. P.W. Barlow & D.J. Carr, pp. 281-318. Cambridge University Press.

Blackwell, P.S. & Wells, E.A. (1983). Limiting oxygen flux densities for oat root extension. *Plant and Soil*, 73, 129-39.

Jackson, M.J. & Drew, M.C. (1984). Effects of flooding on growth and metabolism of herbaceous plants. In *Flooding and Plant Growth*, ed. T.T. Kozlowski, pp. 47-128. London: Academic Press.

Lake, J.V. & Slack, G. (1961). Dependance on light of geotropism in plant roots. *Nature*, 191, 300-1.

Mandoli, D.F. & Briggs, W.R. (1984). Fiber optics in plants. *Scientific American*, 251, 80-8.

Mandoli, D.F., Waldron, L., Nemson, J.A. & Briggs, W.R. (1982). Soil light transmission: implications for phytochrome-mediated responses. *Carnegie Institute of Washington Yearbook*, 1981, 32-4.

Sutton, O.G. (1953). *Micrometeorology*. London: McGraw-Hill.

THE CELLULAR ORGANIZATION OF ROOTS AND ITS RESPONSE TO THE PHYSICAL ENVIRONMENT

P.W. Barlow

INTRODUCTION

All higher plants have roots, though the fraction of the plant's mass that is root varies widely. For example, Spanish moss, *Tillandsia usneoides*, is very nearly rootless in the adult state, whereas members of the Podostemaceae and some orchids (e.g. *Taeniophyllum*) are almost all root when in the vegetative phase. These examples illustrate not only the wide range of forms adopted by roots and root systems (for a review see Barlow, 1986) but also the wide range of environments in which roots exist: *Tillandsia* and *Taeniophyllum* both occupy arid aerial environments, the Podostemaceae live only in the turbulent waters of cataracts.

Although roots thrive in such diverse environments, they are all (with few exceptions) cylindrical and arise in a similar way from an apical meristem. Moreover, the meristem of each species has its own characteristic and, within certain limits, invariant pattern of cellular organization.

Roots may be subjected to great changes in the external environment during their growth. In the soil they encounter variations in temperature, chemical composition and compaction, while aerial and aquatic roots may also experience fluctuations in light quality and quantity. Although many of these environmental variables may severely affect processes in meristematic cells, the root as a whole can accommodate and survive them. Roots are thus strongly homeostatic. This chapter surveys the effects of environmental variables on the cellular processes of root development and asks how perturbations in these processes may be compensated for or corrected.

PROCESSES OF CELL DIVISION

Cell division is cyclical: a cell grows, divides into two, and then both of the two new cells in their turn grow and divide. Within the cell, the nucleus and cytoplasmic organelles also reproduce and divide. Details of cell and nuclear cycles are discussed below, together with effects that environmental factors have upon them. It should be noted that the environmental changes mentioned in this chapter are often imposed by experimentation in the laboratory and are usually applied abruptly. Gradual changes (e.g. in temperature) are much more likely to occur in natural conditions and in these circumstances there is the possibility of some concomitant adaptation.

The cytoskeleton of meristematic cells

For many years cytologists have considered that the interior of the cell is ordered, but only recently have the elements conferring order been identified. The term 'cytoskeleton' is used to denote the macromolecular framework that underpins cell structure and upon which the processes of division, growth and differentiation rely. Many of the responses of cells to environmental effects may ultimately trace to some change in the cytoskeleton.

An important component of the cytoskeleton is the system of microtubules (MTs), the major structural unit of which is the protein tubulin. MTs are readily seen by electron microscopy, but recently they have been seen with the light microscope (which views whole cells rather than the ultra-thin sections of electron microscopy) after conjugating them with fluorescently labelled antibodies (anti-tubulin)(Wick *et al.*, 1981; Wick & Duniec, 1983, 1984).

In interphase cells, MTs labelled with anti-tubulin are seen under the plasmalemma lying at right angles to the principal direction of cell growth (Fig. 1). These cortical MTs are believed to regulate the orientation of the cellulose microfibrils in growing primary cell walls (Heath & Seagull, 1982). As the nucleus approaches mitosis, the cortical MTs disappear and give place to more localized hoops of MTs known as the pre-prophase band (PPB)(Fig. 2). In roots of onion (*Allium cepa*) the earliest PPB is about 5.6-8.3 μm wide, but later narrows to 1.2-3.5 μm (ca. 10% of the cell length)(Wick & Duniec, 1983). Prophase follows with the disappearance of the PPB and the concomitant construction, perhaps from perinuclear MTs, of the mitotic spindle (Fig. 3) upon which the sister chromatids separate. At anaphase-telophase, when the spindle is disassembling, yet another MT array, the phragmoplast, makes its appearance (Fig. 4). At first it is a set of short MTs lying mid-way between the two groups of separated chromosomes (Wick *et al.*, 1981). Later, the phragmoplast spreads centrifugally as an expanding annulus of MTs in whose wake the new cell wall (or cell plate) is formed. Eventually the phragmoplast reaches the sides of the cell and the plate attaches to its walls. A remarkable feature is that the site of attachment closely corresponds to the zone occupied by the PPB earlier in mitosis (for a review see Gunning, 1982a). Each of these three

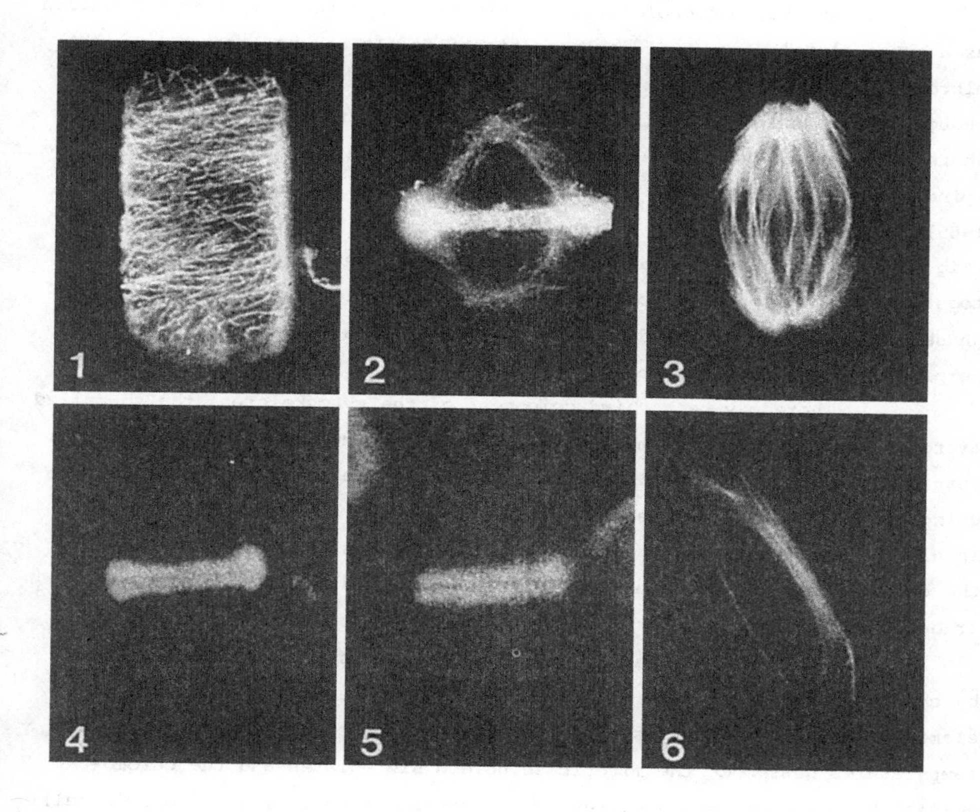

Figs. 1-4. Microtubules (MTs) in meristematic cells of *Allium cepa* revealed by indirect immunofluorescence using anti-tubulin antibody. Fig. 1. Cortical MTs in an interphase cell. Fig. 2. A pre-prophase cell showing a strongly fluorescent hoop of MTs (the PPB) and perinuclear MTs. Fig. 3. MTs of the mitotic spindle of a metaphase cell. Fig. 4. MTs of the phragmoplast of a telophase cell. Figs. 5 and 6. F-actin filaments in cells of *A. cepa* revealed by staining with a fluorescent, rhodamine-labelled derivative of phalloidin. Fig. 5. Actin in the phragmoplast. This is the same cell whose MTs are shown in Fig. 4. Fig. 6. An interphase cell with arrays of actin. Figs. 1 and 2 previously appeared in Clayton & Lloyd (1984) and Figs. 4, 5 and 6 in Clayton & Lloyd (1985); they are reproduced with the permission of the authors, Wissenschaftliche Verlagsgesellschaft and Academic Press.

arrays of MTs - PPB, spindle and phragmoplast - and perhaps the groups of cortical MTs also, are assembled at MT organizing centres (MTOCs); the sequential appearance of these arrays during mitosis suggests a sequential activation of the MTOCs (Lloyd & Barlow, 1982).

Another macromolecule, F-actin, has also recently been recognized as a cytoskeletal element of meristematic cells. It can be located by light microscopy through the use of fluorescently labelled phalloidin, a drug thought to stain actin specifically. Actin is a contractile molecule and has therefore been sought in the mitotic spindle. However, in onion roots both spindle and PPB are free of actin (Clayton & Lloyd, 1985 ; Gunning & Wick, 1985); the only place where MTs and F-actin coexist is the phragmoplast (Fig. 5). Here, both macromolecules appear at the same position and spread together in advance of the cell plate towards the side walls. In some interphase cells F-actin is found running as long fibrils to the inside of, and perpendicular to, the cortical MTs (Fig. 6).

There may be a third component of the cytoskeleton, the so-called microtrabeculae, in meristematic and other cells, but these are not yet well characterised. Microtrabeculae have been revealed in the electron microscope using special preparative techniques (Hawes, Juniper & Horne, 1983; Tiwari *et al.*, 1984). They consist of a lattice of proteins to which are attached the cytoplasmic organelles and perhaps the nucleus also. Such a microtrabecular lattice could serve as a scaffold supporting the MTs and actin.

Any environmental effect upon cell division might be explained by changes in the structure or function of the above-mentioned cytoskeletal elements. One such effect is that of low temperature. MTs depolymerize at temperatures near $0^{\circ}C$, the mitotic spindle disintegrates and the chromosomes, which cannot separate, revert to interphase to give a tetraploid cell; even if the temperature then rises, such cells may be unable to divide.

Many chemicals, including herbicides and fungicides, influence divisions through effects on the MTs; spindle function may be inhibited and cells consequently arrested at metaphase. However, defects of mitosis can also be caused by agents that impair the other cytoskeletal elements involved in division. For example, drugs such as caffeine and deoxyguanosine induce binucleate cells by preventing cell plate formation, apparently by interfering with different stages in the cytoskeletal transformations (Lasselain, 1979). Caffeine is noteworthy because its inhibition of cytokinesis can be mimicked by calcium deficiency (Paul & Goff, 1973) and overcome by addition of Ca^{2+} or Mg^{2+} (Becerra & López-Sáez, 1978). Its effect can also be counteracted by adenosine, but enhanced by dinitrophenol (an uncoupler or oxidative phosphorylation) and low oxygen concentration (4% O_2 in the gas equilibrium phase) (González-Fernández & López-Sáez, 1980; López-Sáez, Mingo & González-Fernández, 1982). These findings suggest that ATP and a Ca^{2+}- or Mg^{2+}-activated ATPase are required for cytokinesis and, together with evidence from ultrastructural studies (López-Sáez, Risueño & Giménez-Martín, 1966; Paul & Goff, 1973), that they operate at the site of the phragmoplast.

Calcium regulates many intracellular processes including polymerization. The protein calmodulin binds Ca^{2+} and by so doing may regulate aspects of both MT and actin activity (Schleicher *et al.*, 1982). Observation of the intracellular location of fluorescently labelled antibodies to calmodulin in relation to the newly forming cortical MTs at early interphase and to the MTs of the PPB in onion and pea (*Pisum sativum*) root meristems gives no indication that calmodulin affects the deployment of MTs at these sites (Wick, Muto & Duniec, 1985). Calmodulin is, however, associated with the spindle and phragmoplast (Wick *et al.*, 1985). Thus, the deployment of calcium-sequestering membranes in the spindle and phragmoplast may influence chromosome separation and cytokinesis. Clearly, any environmental factor that influences the availability of Ca^{2+} and energy sources, or disturbs the integrity of the cytoskeleton, is potentially capable of upsetting mitosis and cell division.

The nuclear cycle

The nuclear cycle is intimately connected with cell growth and division. The two major compartments to the nucleus, the chromosomes and the nucleolus, will be dealt with separately.

Chromosome replication and mitosis

The nuclear cycle provides for the replication of chromosomes, new copies of which are distributed to the daughter cells arising from cytokinesis. The most important environmental variables to which the cycle is susceptible are temperature, oxygen availability and moisture level. Some of the effects of these factors on the replication cycle have been reviewed in detail by Rost (1977).

Nuclei can reproduce and divide over a wide range of temperatures and the duration of the nuclear cycle decreases with increasing temperature (López-Sáez, Giménez-Martín & González-Fernández, 1966; González-Fernández,

Giménez-Martín & De la Torre, 1971). The lower temperature limit for the cycle, e.g. in barley, is about 0.5°C (Grif & Valovich, 1973) and the upper limit is a little above 35°C. Whether mitosis occurs at these limiting temperatures depends on whether the spindle and cytokinetic apparatus are also functional. For instance, nuclei may synthesise DNA and reach mitosis at a low temperature (Barlow & Rathfelder, 1985) but then progress no further because of the failure of tubulin to polymerize and construct a spindle. The optimal temperature for the nuclear cycle varies from species to species. In onion roots the cycle is fastest at 27°C and remains constant above this temperature (Fig. 7); the nuclei can still divide at 35°C but divisions cease at 40°C. Maize (*Zea mays*) shows an inverse relationship between division rate and temperature up to 35°C (Verma, 1980)(Fig. 8). Most observations have been made on crop species of temperate climates; species adapted to grow in other climatic zones (desert, tundra) may have different optimal and limiting temperatures.

Temperature-related changes in division cycle rate are accompanied by proportional changes in the duration of the component phases G_1, S_1, G_2 and mitosis (González-Fernández *et al.*, 1971). There are probably no functions absolutely specific to G_1 and G_2 since these phases are dispensed with under certain circumstances (Cooper, 1979; Navarrete, Cuadrado & Cánovas, 1983); but what regulates the duration of these phases

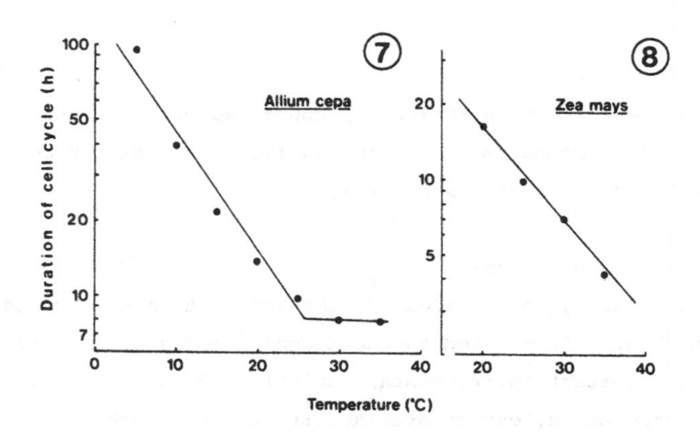

Figs. 7 and 8. Effect of temperature on the duration of the cell cycle in root meristems of *Allium cepa* and *Zea mays*. Data derived from López-Sáez, Giménez-Martín & González-Fernández, (1966) (1966) and Verma, (1980).

when they are present is unknown. Some of the molecular control points regulating the cycle have been reviewed by Giménez-Martín, De la Torre & López-Sáez, 1977): these control points almost certainly correspond to a specific, temporal sequence of gene action, but in plant cells these genes remain unexplored. At the molecular level, events of the S phase are the best characterized. During S, discrete lengths of the DNA double helix (replicons) are replicated in a particular sequence. Temperature influences the rate of movement of the DNA replication fork along the replicon (Van't Hof, Bjerknes & Clinton, 1978). In *Helianthus annuus* the average fork rate is 6 μm/h at 10°C, 8 μm/h at 20°C and 11.5 μm/h at 35°C. Below 15°C the S phase becomes protracted because the initiation of replication is less efficient.

Another temperature-dependent feature of DNA replication is the frequency of occurrence of sister chromatid exchanges (SCEs). In onion roots, exchanges are fewest at 25°C but become more frequent with either warming or cooling (Gutiérrez, Schvartzman & López-Sáez, 1981; Hernández & Gutiérrez, 1983). Gutiérrez *et al.* (1981) and Gutiérrez & López-Sáez (1982) have suggested that the frequency of SCEs may be influenced by intracellular and intranuclear oxygen concentration because in water-grown roots the frequencies are highest at lower temperatures which favour a greater solubility of oxygen. The significance of SCEs is obscure, but they are probably an expression of disturbances in the progress of the DNA replication fork. Because SCEs are the result of chromatid breakage and reunion they are potential sites of mutation, though in a root this is of little direct significance for their function.

Oxygen influences the duration of the nuclear cycle. Interphase in onion roots growing in water bubbled with air (20% oxygen in the gas equilibrium phase) lasts 9.8 h, whereas with 2% oxygen it lasts 41 h (López-Sáez *et al.*, 1969). The effect of oxygen does not become marked until its concentration in the medium falls to below 10%. Of the mitotic phases, metaphase and anaphase are the most sensitive to oxygen deficiency, probably because of the greater oxygen requirement of mitotic spindle assembly and function.

Water potential also influences the nuclear cycle. Exposing roots to different water potentials (e.g. by placing them in varying concentrations of polyethylene glycol (PEG) or mannitol) prolongs the nuclear cycle (González-Bernáldez, López-Sáez & García-Ferrero, 1968; Murín, 1979). This may be the result of a decrease in the rate of turgor-driven growth which, in turn, may directly influence the rate of nuclear progression through the cycle; there could also be direct effects on processes such as DNA synthesis. Yee & Rost (1982) found that mitosis and DNA synthesis in roots of intact bean (*Vicia faba*) plants could recover from a continuous treatment with 20% PEG 6000 (osmotic potential -0.9MPa), but only when the cotyledons were present.

Nucleolar structure

A nucleolar cycle runs in parallel with the chromosome replication cycle (De la Torre & Giménez-Martín, 1982). Early in mitosis the nucleolus disintegrates and its complement of ribonucleoprotein (RNP) is liberated into the cytoplasm, though some remains on the surface of the mitotic chromosomes. When the chromosomes reassemble within a new nuclear membrane at the end of telophase, the nucleolus begins to reform by the coalescence of the RNP material on the chromosomes. The genes for ribosomal RNA (rRNA), which are essential for meristematic cell growth, then become active; the new nucleolus forms at their loci on the chromosomes and continue to grow throughout interphase.

Departures from the cyclical disintegration and reorganization of the nucleolus can be induced by low temperature. In roots of *Allium cepa* (Giménez-Martín *et al.*, 1977), *Brodiaea uniflora* (Sato & Sato, 1984) and *Salix* hybrids (Ehrenberg, 1946) grown at 4-11°C the nucleolus does not disintegrate at prophase but persists into metaphase. A consequence of its persistence is that the nucleolus is freed into the cytoplasm at anaphase. How long it can remain in the cytoplasm, or whether it has any function there, is not known. Nucleolar persistence is also induced by drugs that inhibit, or cause aberrant, RNA (but not protein) synthesis (Moreno Díaz de la Espina, Fernández-Gómez & Risueño, 1979), suggesting that the normal course of nucleolar disintegration at prophase depends on RNA made during the G_2 phase (Fernández-Gómez, Moreno Díaz de la Espina & Risueño, 1978).

Temperature also influences nucleolar size and structure. The volume of nucleoli in root cells of onion grown at 10°C is about twice that of nucleoli at 25°C (Morcillo, Krimer & De la Torre, 1978). Most of this increase results from enlargement of the granular component (Fig. 9) and is accompanied by an increased amount of nucleolar RNA polymerase and a greater abundance of ribosomes in the cytoplasm. However, the over-production of rRNA at the lower temperature is somehow compensated for at a post-transcriptional stage since the amount of protein synthesised per cell division cycle is similar to that at 25°C (De la Torre, Morcillo & Krimer, 1981).

Heat and cold shocks affect nucleolar structure. Six days after transferring young rootlets of maize from 16° to 4°C there is a loss of the granular component, condensation of nucleolus-associated chromatin and the appearance of unusually large (55 nm diameter) RNP particles (Crèvecouer, Deltour & Bronchart, 1983). Transfer of rootlets to 46°C for 3-5 h also disperses the granular component and transforms it into particles 80-140 nm in diameter (Fransolet *et al.*, 1979). Similar results have been obtained for nucleoli of onion transferred from 15°C to 44°C for 3 h (Risueño *et al.*, 1973). A transfer to 35°C is less disruptive, but does cause the appearance of nucleolus-like bodies in the cytoplasm, probably as the result of an upset in the usual nucleolar disintegration/reaggregation events at mitosis (Díez *et al.*, 1971.)

Transfer of onion roots to a near-anoxic environment causes the

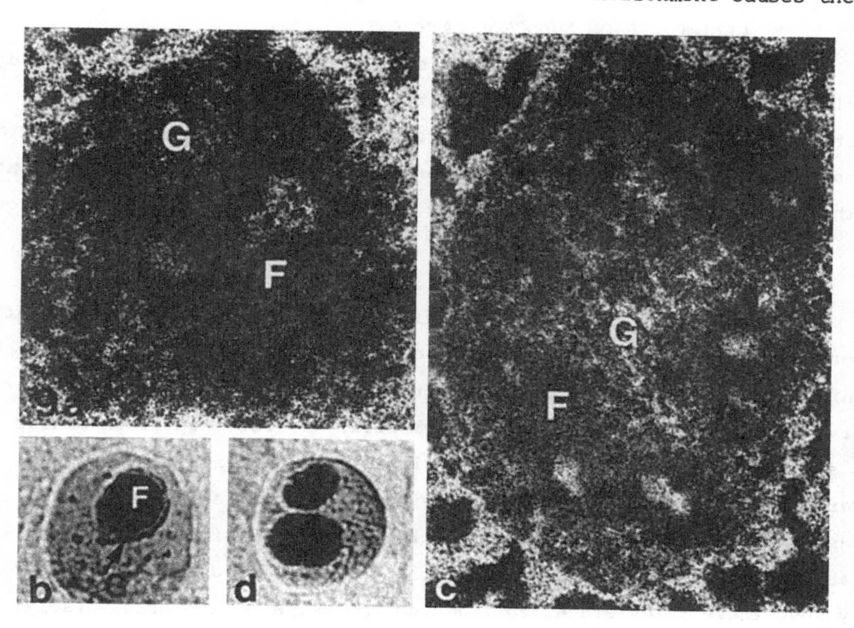

Fig. 9(a,b). Electron and light micrographs showing the segregation of the fibrillar (F) and granular (G) components of a nucleolus in a root meristem cell of *Allium cepa* induced by 3 h of anoxia. (c,d) Electron and light micrographs of a nucleolus in an adequately oxygenated control root showing the typical intermingling of the fibrillar and granular components. Photographs kindly provided by Dr. S. Moreno Díaz de la Espina.

fibrillar component of the nucleolus to segregate from the granular component (Fernández-Gómez, Moreno Díaz de la Espina & Risueño, 1984) (Fig. 9). The effect is complete by 11 h after transfer, but then begins an almost equally complete reversal of segregation. From 18 h a second phase of segregation occurs and is maintained until the roots die at 36 h. The second phase of segregation is completely and rapidly reversible if roots are returned to oxygen-saturated water. Nucleolar segregation is characteristic of treatments that affect rRNA synthesis. The apparent adaptation of the nucleoli to anoxic conditions between the 11th and 18th h is of particular interest since a comparable, though more rapid, adaptation of RNA metabolism has been found in germinating rice embryos transferred from ambient to anoxic conditions (Aspart *et al.*, 1983). Here, after an initial slowing of RNA synthesis, new messenger- and r-RNA molecules are made; they have, however, a different pattern of post-transcriptional processing.

CELL DIVISION AND CELL PATTERNS

The activities of the division cycle described in the preceding section occur in any cell in a root meristem. But meristematic cells do not function in isolation; they are members of a cellular ensemble with a precise organisation and I shall now discuss aspects of division and differentiation as they apply to the generation of the 3-dimensional structure of the root.

Patterns of cell division

It is useful to consider first the relatively simple root of the water-fern, *Azolla pinnata* (Fig. 10) so that the apparent complexity of roots in higher plants can be put in perspective. All cells of the *Azolla* root derive from a single tetrahedral apical cell; the root cap is descended from its acroscopic face, and the three basiscopic descendents of the apical cell undergo a sequence of longitudinal divisions in which each new wall partitions a mother cell at a precise position. These divisions lead to the establishment of every longitudinal file of cells found in the mature root. Only when these files are established do the cells divide transversely. The number of transverse divisions in any file depends on the position of that file within the root. The two types of division, the longitudinal followed by the transverse, are called respectively 'formative' and 'proliferative' (Gunning, Hughes & Hardham, 1978).

The precise positioning of the new cell walls in both formative and proliferative divisions is preceded by the correspondingly precise placements of the PPBs (Gunning, Hardham & Hughes, 1978). No exceptions to this correspondence have been found in undisturbed roots, even though some of the divisions create distinctly unequal pairs of cells. The only exceptions reported follow treatments that interfere with MTs (Gunning, 1982b). Thus, knowledge of whatever regulates the deployment of MTs within the growing cellular ensemble, particularly in the zone of formative divisions, would provide a key to understanding plant morphogenesis.

Although the sequence of formative divisions in the *Azolla* root seems to imply a dauntingly complex control, its deterministic regularity makes it amenable to rational analysis. Lindemayer (1984) has formalized the sequence in terms of a relatively simple control system that relies on transformations of the 'state' of the edges and walls of the cell to specify where new walls will be inserted. The biological counterpart of these 'states' may be the receptivity of zones along the edges of the cell to the PPB.

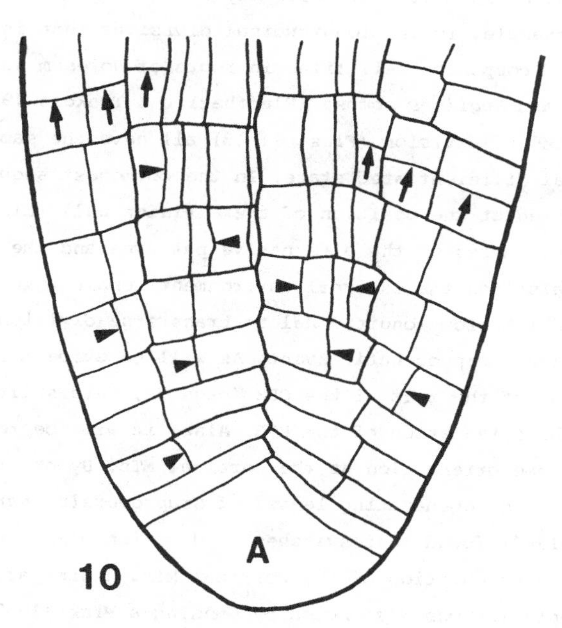

Fig. 10. Longitudinal section through a root apex of *Azolla pinnata* showing the apical cell (A) and its derivatives (root cap not shown). Longitudinal walls arising from recent formative divisions are indicated by darts; transverse walls arising from recent proliferative divisions are indicated by arrows. Redrawn from a micrograph kindly provided by Prof. B.E.S. Gunning.

A major morphological difference between *Azolla* (and some other ferns) and higher plants is that in the latter no cell at the pole of the root functions in a manner comparable to the apical cell. Higher plants have instead a region of slow division rate at the root pole, and the general metabolic quiescence of this region has earned it the name 'quiescent centre' (Clowes, 1956). However, this by no means precludes the quiescent centre from having an important role in root morphogenesis, for it is here that many of the longitudinal (formative) divisions occur. Transverse divisions predominate somewhat proximal to this zone. Again, as in *Azolla*, the number of transverse divisions in a file depends on its position (Luxová, 1975). Whether the formative divisions are as precisely ordered as in *Azolla* has not been analysed in detail, but the predictability of cell patterns at the root pole in different species suggests that they might be (see, for examples, Clowes, 1961; and Seago & Wolniak, 1976). However, a different sequence of divisions in a file of cells may still achieve the same final pattern as, for example, in the longitudinal divisions that initiate the sieve elements and companion cell files in roots of *Hordeum vulgare* (Hagemann, 1957) and *Aegilops comosa* (Eleftheriou & Tsekos, 1982). Three different sequences of division (Figs. 11-13) all have the same outcome in terms of the final differentiated state. In the commonest sequence (Fig. 11) PPBs accurately predict the position of the division wall (Eleftheriou & Tsekos, 1982). The causes of the alternative patterns and their dependence (if any) on signals from the external environment remain unknown.

The shift from longitudinal to transverse divisions in a root is clearly an important morphogenetic event. As will be evident from the earlier discussion of the role of the PPB (page 3), this shift results from a 90° shift in the orientation of the PPB. Also, it will be recalled that the PPB has the same orientation as the cortical MTs. By observing the earliest divisions in regenerating leaves of *Graptopetalum paraguayense*, Selker & Green (1984) found that switches in division plane were accompanied by switches in the orientation of the cortical MTs. A similar switch of cortical MT orientation was also noted by Gunning & Wick (1985) in certain cells of pea roots presumed to be making preparation for longitudinal divisions. The cortical MTs define the orientation of the cellulose microfibrils in the growing cell wall and hence are indicators of the polarity of cell growth (growth is perpendicular to the orientation of the microfibrils). The switch from one plane of division to another might, therefore, be related to a switch in growth polarity.

In a steadily growing root, the pattern of cells is spatially organized. This pattern must have a counterpart in the underlying 3-dimensional pattern of cytoskeletal MTs and wall microfibrils. Because the relative positions of longitudinal and transverse divisions are quite predictable, the orientation of the cytoskeleton must also shift at or near these positions. In the steady state, the position of a cell in an extending file is related to the time elapsed since the initiation of that file from an initial cell. Thus, MT reorientation might result from a time-dependent gene activation; alternatively, it might be a response to a change in the physico-chemical milieu of the cells. Pressure is one determinant of the division plane (Lintilhac & Vesecky, 1984); natural growth regulators are others. Ethylene, for example, can alter MT orientation (Lang, Eisinger & Green, 1981), whereas gibberellin A_3 tends to stabilize MTs against disruption by low temperatures (Mita & Shibaoka, 1984) and so perhaps protects them from factors that would cause their reorientation in its absence. However, whether any of these agents regulate the 3-dimensional pattern of cell divisions in a root meristem is not yet known.

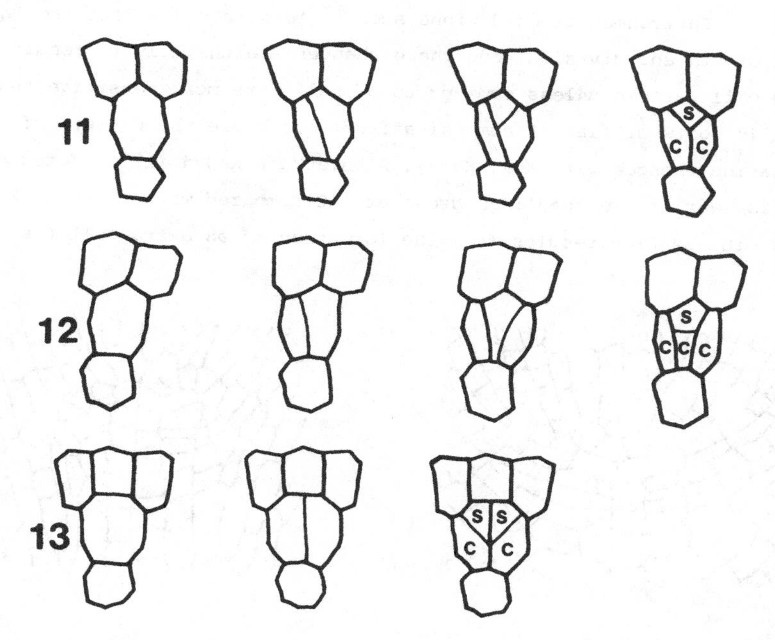

Figs. 11-13. Three alternative sequences of cell division (each should be read from left to right) in the development of protophloem sieve element (S) and companion (C) cells in *Hordeum vulgare*. Redrawn from Hagemann (1957).

Many species conserve the cellular pattern of their roots over long periods. However, roots of some species, e.g. members of the Compositae, show predictable changes of pattern (Armstrong & Heimsch, 1976). The changes consist of a transformation from a 'closed' organisation in which there are distinct tiers of initials for stele, cortex and cap, to an 'open' organisation in which separate initials for cap and cortex are not apparent and the two tissues are continuous. The closed organisation is characteristic of embryonic radicles; it changes to the open organisation early in root growth (Figs. 14 and 15), but later the former closed organisation may be regained. These switches from closed to open and back involve changes in the polarity of cell growth and indicate a corresponding reorientation of the cytoskeleton. A switch in the predominant plane of division in a single tier of cells at the pole of the root can profoundly influence root organisation. For example, if, in an open meristem, all transverse divisions cease in the most proximal tier of the cap, but longitudinal divisions persist, a boundary between the cap and root could become stabilized so converting a meristem to the closed type (Clowes, 1981).

Environmental conditions such as temperature affect the rates of cell growth and division, but these changes are unlikely to result in a changed cell pattern unless some group of cells are more sensitive than others. In roots of flax (*Linum usitatissimum*), where the pattern of cellular organisation changes with age, Kadej, Stobiecka & Kadej (1971) detected this alteration earlier in seedlings grown at 7°C compared with those at 25°C. The new organisation results from the formation of an extra cell tier at the

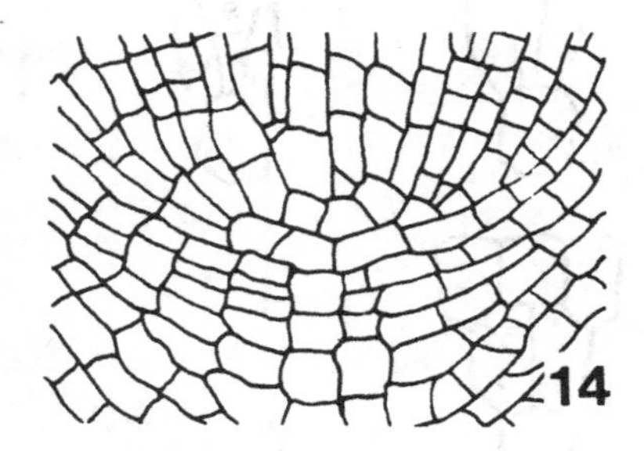

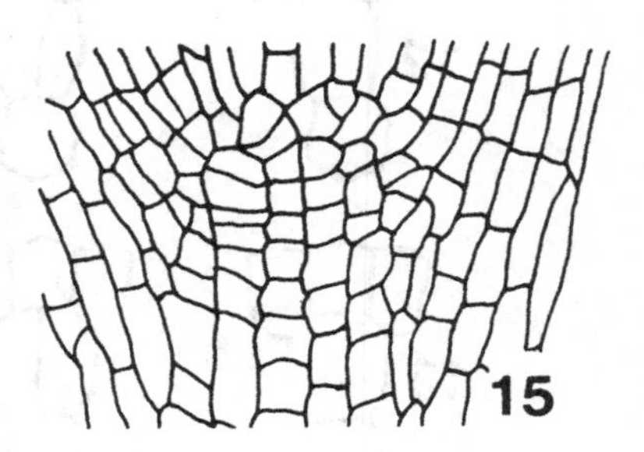

Figs. 14 and 15. Transformation of the initially closed root meristem (Fig. 14) of *Echinacea pallida* (Compositae) to the open type of organization (Fig. 15). Longitudinal sections. Redrawn from Armstrong & Heimsch (1976).

pole of the root as a consequence of a transverse division in a pre-existing tier; this seems to occur earlier at the lower temperature.

Externally applied pressure has been briefly mentioned (page 13) as being able to alter the plane of division. Roots can be subject to pressure when they grow in compacted soil. Pressure, experimentally applied by means of glass ballotini, has been claimed (Wilson & Robards, 1978, 1979) to cause fundamental changes in the cellular organisation of the barley root apex. However, the published photographs and quantitative data supporting this assertion are not convincing. The cell walls are certainly deformed, but their irregularity seem to have misled Wilson & Robards (1979) into believing that the normally closed meristem of barley has become open. Similarly, walls of the endodermis, deformed into oblique orientations (Wilson & Robards, 1978), have probably been misinterpreted as evidence of a pressure-induced stimulation of extra longitudinal divisions in this layer of cells.

The apical organisation in roots of *Selaginella kraussiana* changes when they enter the soil from an aerial environment (Grenville & Peterson, 1981). The aerial root is characterized by a rather blunt tip consisting of a cap with large cells and thick walls (Fig. 16). The soil root is more tapered and typically root-like (Fig. 17). By contrast, aerial roots of *S. martensii* lack a cap, this organ forming when the root enters

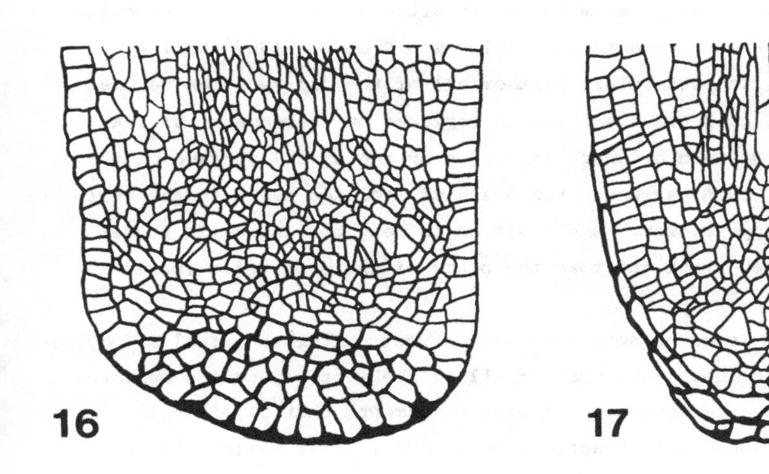

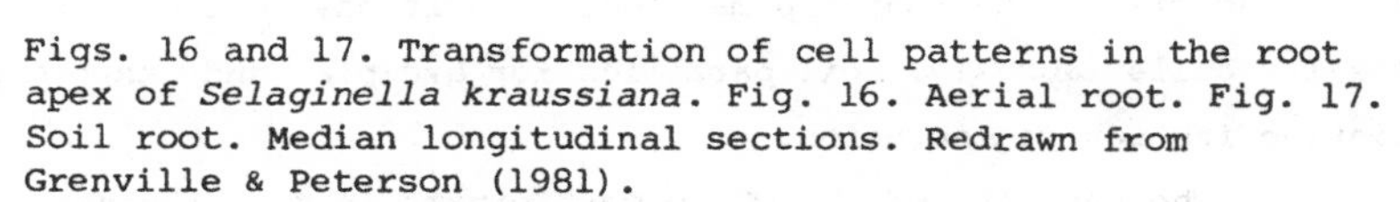

Figs. 16 and 17. Transformation of cell patterns in the root apex of *Selaginella kraussiana*. Fig. 16. Aerial root. Fig. 17. Soil root. Median longitudinal sections. Redrawn from Grenville & Peterson (1981).

the soil or grows in humid air (Webster & Jagels, 1977). The changes of form in roots of both *Selaginella* species result from changes in the pattern of growth and division, particularly in the zone around the initial cell, but the physiological causes are unknown.

Bifurcation is a characteristic of *Selaginella* roots; it is also characteristic of roots of *Pinus* spp., particularly those with an ectomycorrhizal association. It probably results from a temporary interruption of meristematic activity (Wilcox, 1968). When divisions resume, they do so around two new foci of growth which provide the initials for each new meristem. Bifurcation can be induced in non-mycorrhizal roots of *P. radiata* by treatment with the ethylene-releasing compound, 2-chloroethylphosphonic acid; its effectiveness is enhanced by decreased aeration (Wilson & Field, 1984). Thus, both ethylene and poor aeration may interrupt cell growth and division and may also cause the reorientation of the cytoskeleton that eventually leads to root bifurcation.

Regeneration at root apices

The physical environment surrounding a root is likely to vary during its growth. If the environment is too unfavourable, growth may cease but later resume when conditions permit. The resumption of growth is often marked by a discontinuity in the usual activity and organisation of the apex. For example, roots of maize resuming growth after a week at 5°C frequently show changes in meristem organisation (Clowes & Stewart, 1967; Barlow & Rathfelder, 1985). In particular, part or all of the cap meristem becomes inactive after the cold treatment and is replaced by descendents of the quiescent centre (QC) and neighbouring cells at the distal surface of the root proper. Also, cold may cause the death of isolated cells in the meristem. When this happens growth and division in neighbouring cells is reorientated, and they crush and take the place of the dead cells (Barlow & Rathfelder, 1985).

The QC is often activated when a root suffers physical or physiological disturbance. Damage to part or all of the root cap or its meristem (which seems to be hypersensitive to many treatments that impair cell division) is one cause of this activation (Clowes, 1972; Barlow, 1974). Thus, in the field, the root cap may sense conditions that harm it or remove its cells (abrasive soil particles for example) and can respond by initiating its own regeneration.

The capacity of the QC to participate in root regeneration

resides in its ability to remain undamaged by stimuli that damage other meristematic cells. When the QC is activated, its cells and their descendants regenerate an apex similar in construction and pattern of differentiation to the damaged part. The new patterns of division and differentiation depend upon a remodelling not only of the cytoskeleton in each of the activated cells, but also of the internal milieu that causes cells to differentiate in a manner appropriate to their position within the regenerating cellular ensemble (Barlow & Sargent, 1978; Barlow, 1984).

CELL DIFFERENTIATION

Cell differentiation is the result of a particular pattern of gene activity. It is mediated by the cytoplasm, and is expressed in cellular structure and function. The various types of cells in the root differentiate during embryogenesis and once the plan of tissues is established, homoeogenetic induction would seem the most likely means of ensuring their continuity (Barlow, 1984). For this reason, the external physical environment might be expected to have little impact on an already established differentiating system, but would cause only those changes permitted by the intrinsic lability of cellular structure and gene activity.

An example of this lability of structure is the response of tomato (*Lycopersicon esculentum*) roots to sub-optimal oxygen concentrations. Cisternae of the endoplasmic reticulum (ER) in the cortical cytoplasm disappear, starch is lost from the amyloplasts of cortical cells, and mitochondrial structure changes (Morisset, 1978, 1983). The effects are reversible upon re-aeration. In addition, Morisset finds unusually long parallel sheets of ER in the cytoplasm of oxygen-deficient roots. This arrangement of the ER has also been seen in cap, cortical and stelar cells of maize (*Zea mays*) after cold (Mollenhauer, Morré & Vanderwoude, 1975) and water stress (Nir, Klein & Poljakoff-Mayber, 1969; Čiamporová, 1980). Although Mollenhauer *et al*. (1975) interpret this change as evidence of an inhibition of ER membrane flow, another possibility is that it simply represents a rearrangement of ER due to a defect induced in its attachment to the cytoskeleton. It is intriguing that colchicine, which disrupts MTs, also causes multilayered sheets of ER in onion root cells (Witkus, Herold & Vernon, 1982). Moreover, there seems to be a structural interrelationship between MTs and ER (Hensel, 1984) in the central, gravity-perceiving cells of the root cap of cress (*Lepidium sativum*). Here, cold (3-4°C) treatment destroys a class of cortical MTs at the distal end of the cell and permits only feeble development of the distal ER. Returning the roots to warmer (24°C) conditions allows the reformation of MTs and distal ER (Hensel, 1984). Thus, the MTs and ER together form an integrated cytoskeletal system which, in these cells, contributes both to the structural aspect of their differentiated state and perhaps also to their function in the perception of gravity.

In some species further steps in cell differentiation can be induced by environmental factors. For example, in maize, poor aeration triggers the development of air-containing lacunae in the cortex brought about by cell lysis (aerenchyma, Fig. 18). The signal for autolysis, at least in roots grown in poorly aerated culture solution, is ethylene (Drew *et al.* 1981; Konings, 1982), and its effect is further enhanced by nitrogen deficiency (Konings & Verschuren, 1980). Autolysis is probably initiated only when a threshold concentration of the gas within the root is exceeded. Not all cells in the cortex undergo lysis in response to ethylene (Fig. 18); those that do not are possibly either insensitive to, or protected from, the ethylene signal.

Induction of transfer cells in the epidermis of roots is another example of an environmentally regulated pathway of differentiation. Transfer cells are characterized by extensive invaginations of the outer wall with a corresponding increase in the area of plasmalemma. Epidermal transfer cells form when roots of *Atriplex hastata* are grown in saline conditions. However, the same effect is found in roots deprived of an adequate supply of iron. Eleven dicotyledonous species have been examined and all develop epidermal transfer cells in response to iron deficiency; monocotyledonous species dc not respond in this way (Kramer, 1983).

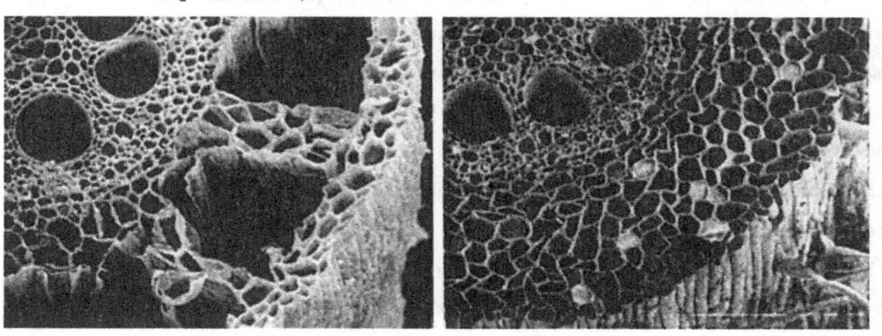

Fig. 18. Aerenchyma in the cortex of a primary root of *Zea mays* formed in response to oxygen deficiency (left). Similar areas are absent in an adequately oxygenated, control root (right). Photographs kindly provided by Dr M.C. Drew.

Changes in cell differentiation accompany the transition from air to soil in roots of *Selaginella kraussiana* (Grenville & Peterson, 1981), *Ficus benghalensis* (Kapil & Rustagi, 1966), and *Monstera deliciosa* (Hinchee, 1981). In the two latter species root hairs develop in soil (but not in air) and the relative amounts of certain tissues (pith, fibres, and vessels in *Ficus*; trichosclereids and raphides in *Monstera*) are altered. In *Selaginella* the most noticeable ultrastructural change is a decrease in the number of cytoplasmic lipid bodies. These are quantitative effects and do not indicate the stimulation of new pathways of differentiation.

Alterations in cell metabolism may mark changes to an already differentiated state. Environmental shocks can result in transient changes in the pattern of protein synthesis that may have no long-lasting effect on cell structure. The new pattern presumably has an adaptive or protective function that lasts for the duration of the shock. For example, maize roots deprived of oxygen synthesise new proteins specific to the anoxic state (Sachs, Freeling & Okimoto, 1980). Whether these proteins are translated from new RNA molecules absent from aerated roots or whether any connection exists between the new proteins and the formation of aerenchyma is unknown. In the experiments of Sachs *et al*. (1980), air in the culture solution was replaced by argon and the roots were unlikely to have formed aerenchyma because the evolution of the extra ethylene necessary for its induction is an oxygen-requiring process (Drew, Jackson & Gifford, 1979).

Analogous to the anoxia proteins are the heat shock proteins made in maize roots in response to raising the temperature from 25°C to 40°C (Cooper & Ho, 1983). Shock proteins are also synthesised by field-grown soyabean (*Glycine max*) when air temperatures approach 40°C (Kimpel & Key, 1985). Some of these proteins may be structural (cytoskeletal) and help protect cytoplasmic organelles from damage during the heat shock phase, dissociating when the temperature falls.

The response of the nucleolus to heat shock was mentioned earlier (page 9). Recent work shows that the synthesis of heat shock proteins and the nucleolar response are related: such proteins can be detected in the nucleolus (Neumann, Scharf & Nover, 1984) and may be related to the unusually large RNP particles that the heat shock induces. Moreover, inhibition of RNA synthesis with actinomycin D suppresses both the heat shock response of the nucleolus and the manufacture of proteins. Developing this work will enable the physiological significance of these heat-induced changes in rRNA processing and protein synthesis to be assessed.

Light has not hitherto been regarded as an important regulator of cellular activity in roots growing in soil. Recently, however, Mandoli & Briggs (1984) have shown that light falling on the shoot of young seedlings can be piped internally through tissue over a distance of at least 2 cm without change in quality. Furthermore, light can penetrate soil for a short distance and be sensed by roots. Light piping through root and soil could allow it to reach the root cap where it could act as a signal for gravitropism in those plants which have this requirement, e.g. some cultivars of maize. A 10 min illumination of maize root caps enhances briefly the level of certain proteins. Although these 'light-shock' proteins turn over within 0.5 h of formation (Feldman & Gildow, 1984), they seem to play a part in the onset of the response to gravity (Feldman, 1983).

FINAL REMARKS

The natural environment of roots is never uniform and roots must accommodate both abrupt and uniform changes in temperature, chemical composition, etc., if they and the shoots they sustain are to survive. Cells are remarkably resilient to injury. However, should cells in the root tip be irreparably damaged other cells can make good their loss enabling the root to maintain its integrity. Homoeostasis of both the structure of individual cells and the pattern of multicellular units is a remarkable feature of roots; homoeostasis probably also applies to the root system as a whole. Future research at the subcellular, cellular and organ levels, particularly using environmental variables as experimental probes, may well reveal the causes of this intrinsic stability of root form.

REFERENCES

Armstrong, J.E. & Heimsch, C. (1976). Ontogenetic reorganization of the root meristem in the Compositae. *American Journal of Botany*, 63, 212-9.

Aspart, L., Got, A., Delseny, M., Mocquot, B. & Pradet, A. (1983). Adaptation of ribonucleic acid metabolism to anoxia in rice embryos. *Plant Physiology*, 72, 115-21.

Barlow, P. (1974). Regeneration of the cap of primary roots of *Zea mays*. *New Phytologist*, 73, 937-54.

Barlow, P.W. (1984). Positional control in root development. In *Positional Controls in Plant Development*, ed. P.W. Barlow & D.J. Carr, pp.281-318. Cambridge University Press.

Barlow, P.W. (1986). Adventitious roots of whole plants: their forms, functions and evolution. In *New Root Formation in Plants and Cuttings*, ed. M.B.Jackson, pp.67-110. The Hague: M. Nijhoff/W. Junk.

Barlow, P.W. & Rathfelder, E.L. (1985). Cell division and regeneration in primary root meristems of *Zea mays* recovering from cold treatment. *Environmental and Experimental Botany*, 25, 303-14.

Barlow, P.W. & Sargent, J.A. (1978). The ultrastructure of the regenerating root cap of *Zea mays* L. *Annals of Botany*, 42, 791-9.

Becerra, J. & López-Sáez, J.F. (1978). Effects of caffeine, calcium and magnesium on plant cytokinesis. *Experimental Cell Research*, 111, 301-8.

Čiamporová, M. (1980) Ultrastructure of cortical cells of maize root under water stress conditions. *Biologia Plantarum*, 22, 444-9.

Clayton, L. & Lloyd, C.W. (1984). The relationship between the division plane and spindle geometry in *Allium* cells treated with CIPC and griseofulvin: an anti-tubulin study. *European Journal of Cell Biology*, 34, 248-53.

Clayton, L. & Lloyd, C.W. (1985). Actin organisation during the cell cycle in meristematic plant cells. Actin is present in the cytokinetic phragmoplast. *Experimental Cell Research*, 156, 231-8.

Clowes, F.A.L. (1956). Localization of nucleic acid synthesis in root meristems. *Journal of Experimental Botany*, 7, 307-12.

Clowes, F.A.L. (1961). *Apical Meristems*. Oxford: Blackwell.

Clowes, F.A.L. & Stewart, H.E., (1967). Recovery of dormancy in roots. *New Phytologist*, 66, 115-23.

Clowes, F.A.L. (1972). Regulation of mitosis in roots by their caps. *Nature New Biology*, 235, 143-4.

Clowes, F.A.L. (1981). The difference between open and closed meristems. *Annals of Botany*, 48, 761-7.

Cooper, P. & Ho, T.-H.D. (1983). Heat shock proteins in maize. *Plant Physiology*, 71, 215-22.

Cooper, S. (1979). A unifying model for the G1 period in prokaryotes and eukaryotes. *Nature*, 280, 17-9.

Crèvecoeur, M., Deltour, R. & Bronchart, R. (1983). Effects of subminimal temperature on physiology and ultrastructure of *Zea mays* embryo during germination. *Canadian Journal of Botany*, 61, 1117-25.

De la Torre, C. & Giménez-Martín, G. (1982). The nucleolar cycle. *Society for Experimental Biology Seminar Series*, 15, 153-77.

De la Torre, C., Morcillo, G. & Krimer, D.B. (1981). Coupling of replication, transcription and translation under two different steady state conditions in *Allium cepa* meristems. *Cytobios*, 30, 7-18.

Díez, J.L., Marín, M.F., Esponda, P. & Stockert, J.C. (1971). Prenucleolar bodies in the cytoplasm of meristematic cells after thermal shock. *Experientia*, 27, 266-7.

Drew, M.C., Jackson, M.B. & Giffard, S. (1979). Ethylene-promoted adventitious rooting and development of cortical air spaces (aerenchyma) in roots may be adaptive responses to flooding in *Zea mays* L. *Planta*, 147, 83-8.

Drew, M.C., Jackson, M.B., Giffard, S.C. & Campbell, R. (1981). Inhibition by silver ions of gas space (aerenchyma) formation in adventitious roots of *Zea mays* L. subjected to exogenous ethylene or to oxygen deficiency. *Planta*, 153, 317-24.

Ehrenberg, L. (1946). Influence of temperature on the nucleolus and its coacervate nature. *Hereditas*, 32, 407-18.

Eleftheriou, E.P. & Tsekos, I. (1982). Development of protophloem in roots of *Aegilops comosa* var. *thessalica*. I. Differential divisions and pre-prophase bands of microtubules. *Protoplasma*, 113, 110-9.

Feldman, L.J. (1983). Light-enhanced protein synthesis in gravitropically stimulated root caps of corn. *Plant Physiology*, 72, 833-6.

Feldman, L.J. & Gildow, V. (1984). Effects of light on protein pattern in gravitropically stimulated root caps of corn. *Plant Physiology*, 74, 208-12.

Fernández-Gómez, M.E., Moreno Díaz de la Espina, S. & Risueño, M.C. (1978). Alteration of nucleolar dispersion in prophase by 5-FUdR treatment. *Cell Biology International Reports*, 2, 237-44.

Fernández-Gómez, M.E., Moreno Díaz de la Espina, S. & Risueño, M.C. (1984). Nucleolar activity in anoxic root meristem cells of *Allium cepa*. *Environmental and Experimental Botany*, 24, 219-28.

Fransolet, S., Deltour, R., Bronchart, R. & Van de Walle, C. (1979). Changes in ultrastructure and transcription induced by elevated temperature in *Zea mays* embryonic root cells. *Planta*, 146, 7-18.

Giménez-Martín, G., De la Torre, C. & López-Sáez, J.F. (1977). Cell division in higher plants. In *Mechanisms and Control of Cell Division*, ed. T.L. Rost & E.M. Gifford, Jr., pp. 261-307. Stroudsburg: Dowden, Hutchinson & Ross.

Giménez-Martín, G., De la Torre, C., López-Sáez, J.F. & Esponda, P. (1977). Plant nucleolus: structure and physiology. *Cytobiologie*, 14, 421-62.

González-Bernáldez, F., López-Sáez, J.F. & Garcia-Ferrero, G. (1968). Effect of osmotic pressure on root growth, cell cycle and cell elongation. *Protoplasma*, 65, 255-62.

González-Fernández, A., Giménez-Martín, G. & De la Torre, C. (1971). The duration of the interphase periods at different temperatures in root tip cells. *Cytobiologie*, 3, 367-71.

González-Fernández, A. & López-Sáez, J.F. (1980). The effect of adenosine on caffeine inhibition of plant cytokinesis. *Environmental and Experimental Botany*, 20, 455-61.

Grenville, D.J. & Peterson, R.L. (1981). Structure of aerial and subterranean roots of *Selaginella kraussiana* A. Br. *Botanical Gazette*, 142, 73-81.

Grif, V.G. & Valovich, E.M. (1973). The mitotic cycle of plant cells at the minimal temperature of mitosis. *Tsitologiya*, 15, 1510-4 (in Russian).

Gunning, B.E.S. (1982a). The cytokinetic apparatus: its development and spatial regulation. In *The Cytoskeleton in Plant Growth and Development*, ed. C.W. Lloyd, pp. 229-92. London: Academic Press.

Gunning, B.E.S. (1982b). The root of the water fern *Azolla:* cellular basis of development and multiple roles for cortical microtubules. In *Developmental Order: Its Origin and Regulation*, ed. S. Subtelny & P.B. Green, pp. 379-421. New York: Liss.

Gunning, B.E.S., Hardham, A.R. & Hughes, J.E. (1978). Pre-prophase bands of microtubules in all categories of formative and proliferative cell division in *Azolla* roots. *Planta*, 143, 145-60.

Gunning, B.E.S., Hughes, J.E. & Hardham, A.R. (1978). Formative and proliferative cell divisions, cell differentiation, and developmental changes in the meristem of *Azolla* roots. *Planta*, 143, 121-44.

Gunning, B.E.S. & Wick, S.M. (1985). Preprophase bands, phragmoplasts, and spatial control of cytokinesis. *Journal of Cell Science, Supplement*, 2, 157-79.

Gutiérrez, C. & López-Sáez, J.F. (1982). Oxygen dependence of sister-chromatid exchanges. *Mutation Research*, 103, 295-302.
Gutiérrez, C., Schvartzman, J.B. & López-Sáez, J.F. (1981). Effect of growth temperature on the formation of sister-chromatid exchanges in BrdUrd-substituted chromosomes. *Experimental Cell Research*, 134, 73-79.
Hagemann, R. (1957). Anatomische Untersuchungen an Gerstenwurzeln. *Die Kulturpflanze*, 5, 75-107.
Hawes, C.R., Juniper, B.E. & Horne, J.C. (1983). Electron microscopy of resin-free sections of plant cells. *Protoplasma*, 115, 88-93.
Heath, I.B. & Seagull, R.W. (1982). Oriented cellulose fibrils and the cytoskeleton: a critical comparison of models. In *The Cytoskeleton in Plant Growth and Development*, ed. C.W. Lloyd, pp. 163-82. London: Academic Press.
Hensel, W. (1984). A role of microtubules in the polarity of statocytes from roots of *Lepidium sativum* L. *Planta*, 162, 404-14.
Hernández, P. & Gutiérrez, C. (1983). Sister-chromatid exchange formation at supra-optimal growth temperatures. *Mutation Research*, 108, 293-9.
Hinchee, M.A.W. (1981). Morphogenesis of aerial and subterranean roots of *Monstera deliciosa*. *Botanical Gazette*, 142, 347-59.
Kadej, A., Stobiecka, H. & Kadej, F. (1971). Organisation of the root apical meristem in *Linum usitatissimum* L. grown at 25°C and 7°C. *Acta Societatis Botanicorum Poloniae*, 40, 389-94.
Kapil, R.N. & Rustagi, P.N. (1966). Anatomy of the aerial and terrestrial roots of *Ficus benghalensis L. Phytomorphology*, *16*, 382-6.
Kimpel, J.A. & Key, J.L. (1985). Heat shock in plants. *Trends in Biochemical Sciences*, 10, 353-57.
Konings, H. (1982). Ethylene-promoted formation of aerenchyma in seedling roots of *Zea mays* L. under aerated and non-aerated conditions. *Physiologia Plantarum*, 54, 119-24.
Konings, H. & Verschuren, G. (1980). Formation of aerenchyma in roots of *Zea mays* in aerated solutions, and its relation to nutrient supply. *Physiologia Plantarum*, 49, 265-70.
Kramer, D. (1983). Genetically determined adaptations in roots to nutritional stress: correlation of structure and function. *Plant and Soil*, 72, 167-73.
Lang, J.M., Eisinger, W.R. & Green, P.B. (1981). Effects of ethylene on the orientation of microtubules and cellulose microfibrils of pea epicotyl cells with polylamellate cell walls. *Protoplasma*, 110, 5-14.
Lasselain, M.-J. (1979). Etude ultrastructurale comparative de l'action inhibitrice de la désoxyguanosine et de celle de la caféine sur la cytodiérèse d'*Allium sativum* *Comptes Rendu des Séances de la Société de Biologie*, 173, 26-32.
Lintilhac, P.M. & Vesecky, T.B. (1984). Stress-induced alignment of division plane in plant tissues grown *in vitro*. *Nature*, 307, 363-4.
Lindenmayer, A. (1984). Models for plant tissue development with cell division orientation regulated by preprophase bands of microtubules. *Differentiation*, 26, 1-10.
Lloyd, C.W. & Barlow, P.W. (1982). The co-ordination of cell division and elongation: the role of the cytoskeleton. In *The Cytoskeleton in Plant Growth and Development*, ed. C.W. Lloyd, pp. 203-28. London: Academic Press.

López-Sáez, J.F., Giménez-Martín, G. & González-Fernández, A. (1966). Duration of the cell division cycle and its dependence on temperature. *Zeitschrift für Zellforschung*, 75, 591-600.
López-Sáez, J.F., González-Bernáldez, F., González-Fernández, A. & García-Ferrero, G. (1969). Effect of temperature and oxygen tension on root growth, cell cycle and cell elongation. *Protoplasma*, 67, 213-21.
López-Sáez, J.F., Mingo, R. & González-Fernández, A. (1982). ATP level and caffeine efficiency on cytokinesis inhibition in plants. *European Journal of Cell Biology*, 27, 185-90.
López-Sáez, J.F., Risueño, M.C. & Giménez-Martín, G. (1966). Inhibition of cytokinesis in plant cells. *Journal of Ultrastructure Research*, 14, 85-94.
Luxová, M. (1975). Some aspects of the differentiation of primary root tissues. In *The Development and Function of Roots*, ed. J.G. Torrey & D.T. Clarkson, pp. 73-90. London: Academic Press.
Mandoli, D.F. & Briggs, W.R. (1984). Fiber optics in plants. *Scientific American*, 251, 80-8.
Mita, T. & Shibaoka, H. (1984). Gibberellin stabilizes microtubules in onion leaf sheath cells. *Protoplasma*, 119, 100-9.
Mollenhauer, H.H., Morré, D.J., & Vanderwoude, W.J. (1975). Endoplasmic reticulum-Golgi apparatus associations in maize root tips. *Mikroskopie*, 31, 257-72.
Morcillo, G., Krimer, D.B. & De la Torre, C. (1978). Modification of nucleolar components by growth temperature in meristems. *Experimental Cell Research*, 115, 95-102.
Moreno Díaz de la Espina, S., Fernández-Gómez, M.E. & Risueño, M.C. (1979). Occurrence of nucleolar material in the cytoplasm of plant cells. *Cell Biology International Reports*, 3, 215-25.
Morisset, C. (1978). Structural and cytoenzymological aspects of the mitochondria in excised roots of oxygen-deprived *Lycopersicum* cultivated *in vitro*. In *Plant Life in Anaerobic Environments*, ed. D.D. Hook & R.M.M. Crawford, pp. 497-537. Ann Arbor: Ann Arbor Science Publishers.
Morisset, C. (1983). Effects of energetic shortage upon the ultrastructure of some organelles, in excised roots of *Lycopersicum esculentum* cultivated in vitro I. Reversible structural modifications of the endoplasmic reticulum. *Cytologia*, 48, 349-62.
Murín, A. (1979). Effects of high osmotic potential of a medium on mitotic cycle in roots of *Vicia faba* L. *Biologia Plantarum*, 21, 345-50.
Navarrete, M.H., Cuadrado, A. & Cánovas, J.L. (1983). Partial elimination of G_1 and G_2 periods in higher plant cells by increasing the S period. *Experimental Cell Research*, 148, 273-80.
Neumann, D., Scharf, K.-D. & Nover, L. (1984). Heat shock induced changes of plant cell ultrastructure and autoradiographic localization of heat shock proteins. *European Journal of Cell Biology*, 34, 254-64.
Nir, I., Klein, S., & Poljakoff-Mayber, A. (1969). Effect of moisture stress on submicroscopic structure of maize roots. *Australian Journal of Biological Sciences*, 22, 17-33.
Paul, D.C. & Goff, C.W. (1973). Comparative effects of caffeine, its analogues and calcium deficiency on cytokinesis. *Experimental Cell Research*, 78, 399-413.
Risueño, M.C., Stockert, J.C., Giménez-Martín, G. & Díez, J.L. (1973). Effect of supraoptimal temperatures on meristematic cells nucleoli. *Journal of Microscopie*, 16, 87-94.

Rost, T.L. (1977). Responses of the plant cell cycle to stress. In *Mechanisms and Control of Cell Division*, ed. T.L. Rost & E.M. Gifford, Jr., pp. 111-43. Stroudsburg: Dowden, Hutchinson & Ross.

Sachs, M.M., Freeling, M. & Okimoto, R. (1980). The anaerobic proteins of maize. *Cell*, 20, 761-7.

Sato, S. & Sato, M. (1984). Peculiar behavior of the nucleolus and appearance of cytoplasmic nucleolus-like bodies in the root tip meristems of *Brodiaea uniflora* Engl. grown at low temperature. *Protoplasma*, 120, 197-208.

Schleicher, M., Everson, D.B., Van Eldik, L.J. & Watterson, D.M. (1982). Calmodulin. In *The Cytoskeleton in Plant Growth and Development*, ed. C.W. Lloyd, pp. 85-106. London: Academic Press.

Seago, J.L. & Wolniak, S.M. (1976). Cortical ontogeny in roots. I. *Zea mays*. *American Journal of Botany*, 63, 220-5.

Selker, J.M.L. & Green, P.B. (1984). Organogenesis in *Graptopetalum paraguayense* E. Walther: shifts in orientation of cortical microtubule arrays are associated with periclinal divisions. *Planta*, 160, 289-97.

Tiwari, S.C., Wick, S.M., Williamson, R.E. & Gunning, B.E.S. (1984). Cytoskeleton and integration of cellular function in cells of higher plants. *The Journal of Cell Biology*, 99, 63s-9s.

Van't Hof, J., Bjerknes, C.A. & Clinton, J.H. (1978). Replicon properties of chromosomal DNA fibers and the duration of DNA synthesis of sunflower root-tip meristem cells at different temperatures. *Chromosoma*, 66, 161-71.

Verma, R.S. (1980). The duration of G_1, S, G_2, and mitosis at four different temperatures in *Zea mays* L. as measured with ^{3}H-thymidine. *Cytologia*, 45, 327-33.

Webster, T.R. & Jagels, R. (1977). Morphology and development of aerial roots of *Selaginella martensii* grown in moist containers. *Canadian Journal of Botany*, 55, 2149-58.

Wick, S.M. & Duniec, J. (1983). Immunofluorescence microscopy of tubulin and microtubule arrays in plant cells. I. Preprophase band development and concomitant appearance of nuclear envelope-associated tubulin. *The Journal of Cell Biology*, 97, 235-43.

Wick, S.M. & Duniec, J. (1984). Immunofluorescence microscopy of tubulin and microtubule arrays in plant cells. II. Transition between the pre-prophase band and mitotic spindle. *Protoplasma*, 122, 45-55.

Wick, S.M., Muto, S. & Duniec, J. (1985). Double immunofluorescence labelling of calmodulin and tubulin in dividing plant cells. *Protoplasma*, 126, 198-206.

Wick, S.M., Seagull, R.W., Osborn, M., Weber, K. & Gunning, B.E.S. (1981). Immunofluorescence microscopy of organized microtubule arrays in structurally stabilized meristematic plant cells. *The Journal of Cell Biology*, 89, 685-90.

Wilcox, H.E. (1968). Morphological studies of the roots of red pine, *Pinus resinosa*. II. Fungal colonization of roots and the development of mycorrhizae. *American Journal of Botany*, 55, 688-700.

Wilson, A.J. & Robards, A.W. (1978). The ultrastructural development of mechanically impeded barley roots. Effects on the endodermis and pericycle. *Protoplasma*, 95, 255-65.

Wilson, A.J. & Robards, A.W. (1979). Some observations of the effects of mechanical impedance upon the ultrastructure of the root cap of barley. *Protoplasma*, 101, 61-72.

Wilson, E.R.L. & Field, R.J. (1984). Dichotomous branching in lateral roots of pine: the effect of 2-chloroethylphosphonic acid on seedlings of *Pinus radiata* D. Don. *New Phytologist*, 98, 465-73.

Witkus, R., Herold, L. & Vernon, G.M. (1982). An unusual arrangement of smooth membranes in the root meristem cells of *Allium cepa* L. following exposure to colchicine. *Experimental Cell Biology*, 50, 56-8.

Yee, V.F. & Rost, T.L. (1982). Polyethylene glycol induced water stress in *Vicia faba* seedlings: cell division, DNA synthesis and a possible role for cotyledons. *Cytologia*, 47, 615-24.

FUNCTIONAL ASPECTS OF CELLS IN ROOT APICES

William J. Lucas

INTRODUCTION

To survive and function within the physically harsh environment of the soil, roots of terrestrial plants have developed a range of important anatomical and physiological features. The root system of a particular plant species may well encounter a range of soils, all with differing structural and chemical properties as well as biotic composition, thus the root system must also be capable of adapting to these changes in soil properties. In this chapter we shall examine selected examples to illustrate both the function of these cells within the root apex and their interactive nature.

ACQUISITION OF MINERAL NUTRIENTS

Although cells of the root apex can take up macronutrients like K^+, Cl^-, etc., from the soil solution (Torii & Laties, 1966), these ions are generally supplied to the shoot by cells that are much further removed from the apex. On the other hand, certain essential micronutrients like Fe seem to be taken up by cells within the first few millimeters of the root tip (Ambler, Brown & Gauch, 1971; Brown & Ambler, 1974; Clarkson & Sanderson, 1978; Römheld & Marschner, 1981b). We will use the example of Fe nutrition to illustrate the point that in many plants, the cells of the root apex have the capability to respond to changes in the levels of available nutrient ions within the microenvironment of the soil. At the outset, we must stress that not all plants respond in the same way when challenged with Fe deficiency.

When an Fe-efficient plant like *Helianthus* (sunflower) is transferred to Fe-deficient culture conditions, the development of the root apex is dramatically changed. The cortex thickens considerably and root hair development is enhanced in comparison with that in plants supplied with adequate Fe. Kramer *et al.* (1980) conducted a cytological investigation on *Helianthus* roots obtained from plants that had been transferred to nutrient solutions free of Fe. They found that within 24-48 hours of Fe deficiency most of the peripheral cells within the root tips had differentiated into transfer cells (for a description of the characteristics of these cells, see Gunning, 1977). The interesting feature of this cytological response was that the wall labyrinths that developed were always located on the peripheral walls facing the external medium. Kramer *et al.* (1980) proposed that these rhizodermal transfer cells are in some way responsible for the Fe-efficient response of sunflower.

Kramer (1983) and Römheld & Kramer (1983) further investigated the role of rhizodermal transfer cells in terms of a response to Fe deficiency. They found that Fe deficiency-induced transfer cells were present in all dicotyledonous species examined, but in the monocots, only two *Allium* (onion) species developed these characteristics; in these species the wall labyrinths were quite limited in comparison to sunflower. Iron-efficient plants, like *Zea mays* (maize), did not develop transfer cells in their root apices. These studies provide the basis for interesting future research, especially in terms of elucidating the genetic determinant for, and the basic physiological mechanism by which these transfer cells are induced.

Other physiological processes that are also influenced by Fe deficiency include an enhancement in both the rate of net H^+ efflux from the root apex (Römheld & Marschner, 1981b), and the capability of the root tip to reduce Fe^{3+} (Bienfait, van der Bliek & Bino, 1982; Marschner, Römheld & Ossenberg-Neuhaus, 1982). Both processes may increase the supply of available Fe to the root, and that in association with the rhizodermal transfer cells, could account for the enhanced capacity for Fe absorption, a phenomenon observed in Fe-efficient plants (see, for example, Römheld & Marschner, 1981a).

Experimental data consistent with the presence of a regulatory mechanism involved in Fe uptake (by the root tips) were obtained by Römheld & Marschner (1981a). They used both hydroponic- and soil-cultured sunflower plants grown under *suboptimal* Fe conditions. As illustrated in Figs. 1 and 2, they detected a distinct rhythm in ^{59}Fe uptake under both culture conditions. The rhythmic changes in ^{59}Fe occurred every 2-4 days, simultaneously in roots, shoots and in the exudate collected from the excised sunflower stems (Fig. 1). The rhythm in ^{59}Fe uptake observed in plants grown in calcareous soil (Fig. 2) is very important because it represents both ecological and agronomic situations where the soil is low in available Fe. Based on these studies, Römheld & Marschner (1981a) suggested that, "the iron nutritional status of the sunflower plant is transformed into a 'signal' which induces distinct biochemical and morphological changes within the root tips and causes a fine regulation of iron supply to the plant." A general scheme of this regulation of Fe uptake is presented in Fig. 3.

In an integrated ultrastructural and physiological study of the roots of *Helianthus annuus, Cucumis sativus, Solanum tuberosum, Allium porrum, Coleus blumei* and *Zea mays*, Römheld & Kramer (1983) failed to find the expected correlation between the induction of transfer cells and the

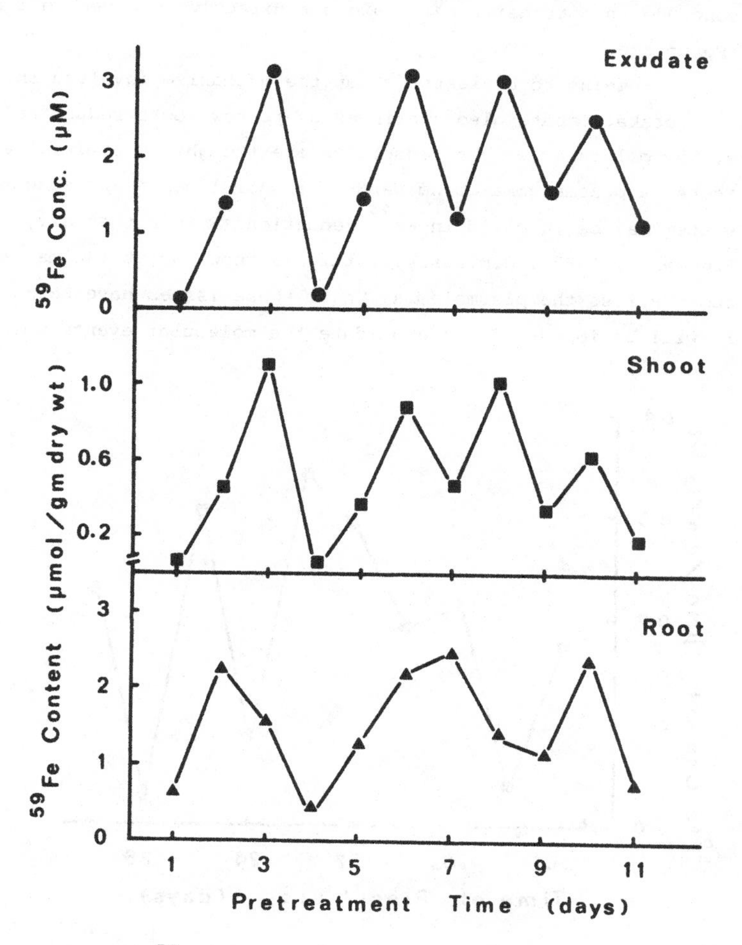

Figure 1. ^{59}Fe content in roots, shoots and exudate of hydroponically-grown sunflower plants (*Helianthus annuus*) pretreated with 2 x 10^{-6} M ferric ethylenediaminedi(O-hydroxyphenylacetic acid)(FeEDDHA unlabelled) for 1 to 11 days. Every day two plants were supplied with 2x10^{-5} M ^{59}FeEDDH . (Redrawn from Römheld & Marschner, 1981a).

enhanced ability both to reduce Fe^{3+} and secrete protons. What they did find was a good correlation between the formation (and extent) of rhizodermal transfer cells and the rate of net proton efflux induced by Fe deficiency. The poor correlation between the observed increased capacity for Fe absorption and the extent of rhizodermal transfer cell development is perplexing. Until more information is available about the properties of the wall labyrinth that develops in the transfer cells of different species, we cannot conclude that transfer cells are not directly involved in the process of Fe uptake.

Much remains to be learnt about the mechanism involved in regulating Fe uptake. Debate also continues as to how roots reduce Fe^{3+}. In some plants, low molecular weight reductants are thought to be involved, while in others, a plasmalemma-bound NADH- (or NADPH) dependent electron transport system may be involved in Fe^{3+} reduction (Beinfait *et al.*, 1983); Sijmons & Bienfaith, 1983). Similarly, little is known about the mechanism of Fe transport across the plasmalemma. Until these issues have been resolved, it will be impossible to elucidate the molecular events which

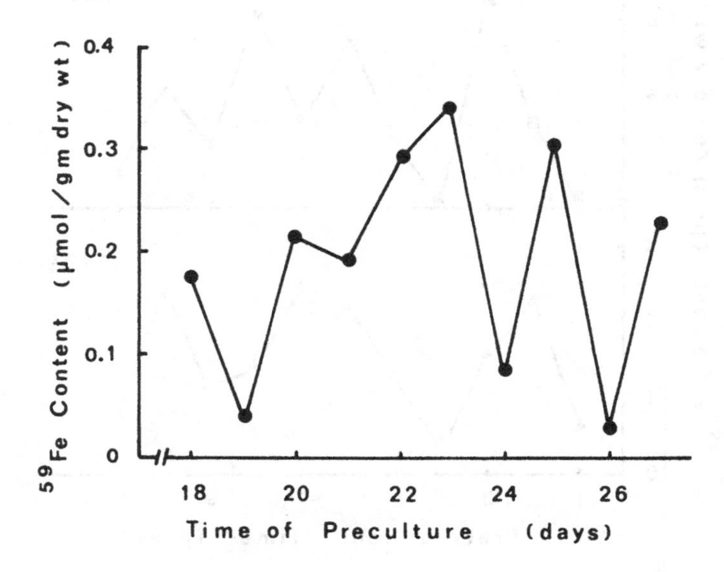

Figure 2. ^{59}Fe content in shoots of sunflower plants precultured on calcareous soil for 18 to 27 days. After different times $^{59}FeEDDHA$ (2×10^{-5}M) was percolated through the pots and the plants harvested after 24 hours. (Redrawn from Römheld & Marschner, 1981a).

regulate the iron nutritional status of a plant.

ROOT PENETRATION

Soil physical properties: influence on root growth

It is common knowledge among agriculturalists that the physical characteristics of the soil influence the development of roots, modifying both the growth rate and the distribution within the soil. Root penetration through the soil may be greatly influenced by the size of the pores between the soil particles and the degree of compaction (i.e., compacted soils would need greater pressures to enlarge the soil pores). Thus, the physical

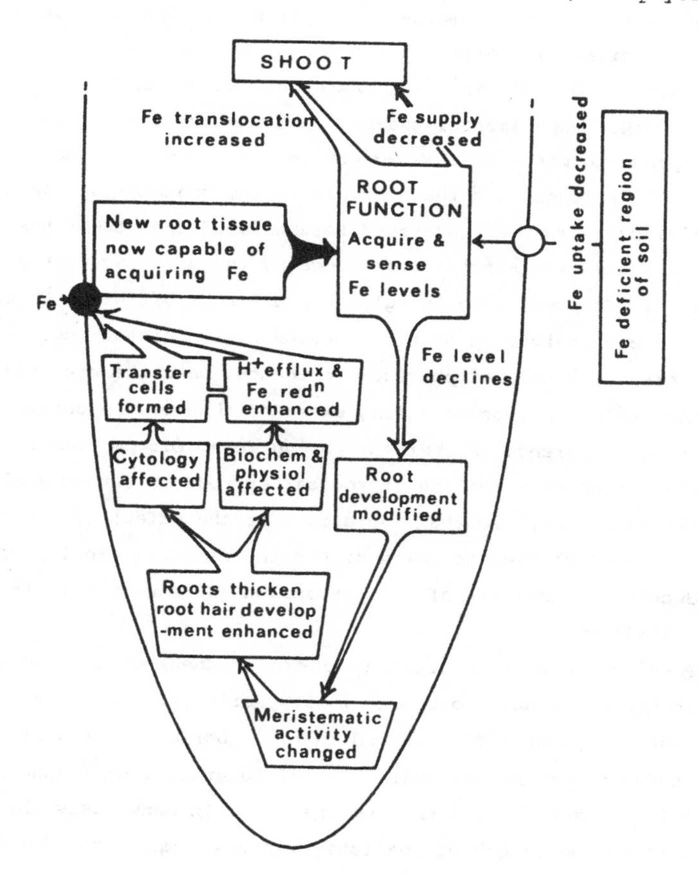

Figure 3. Schematic representation of the dicot root response to changing levels of available Fe in the soil. The mechanism by which the tissue of the root tip actually 'senses' the change in tissue Fe is yet to be elucidated.

properties of the soil can modify root diameter, development of root hairs and the branching pattern of lateral roots. These responses show some similarities to those observed in plants responding to Fe deficiency; again the site of perception and response is thought to be located in the root tip (see Feldman, 1984 and references therein).

Numerous workers have studied the relationship between the characteristics of the soil and the response of various plant root systems. Goss and coworkers used an ingenious experimental chamber which allowed them to vary, independently, the pore size and the applied pressure within the root zone. With this system, Goss (1977) found that in many agriculturally important crops (barley, wheat, maize and sugar beet), the rate of root elongation was considerably decreased (to about one-half) by rather small mechanical stresses (20 to 50 kPa). An anatomical study conducted on barley plants, grown in this experimental chamber, provided valuable details on the root's response to mechanical impedance (Wilson, Robards & Goss, 1977). Wilson *et al.* (1977) found that the increase in the diameter of the barley roots resulted largely from an increased thickness of the cotex; the number of cells in the transverse section increased and the diameter of the outer cells was greater. However, surprisingly, the diameter of the inner cells was decreased. The organization of the vascular tissue of the root was also modified by these low levels of physical impedance. Near the apex the diameter of the stele, in impeded roots, exceeded that of the control roots, with the greatest difference (22 percent) being found 0.5 cm from the apex.

These complex reactions of the barley root to rather small physical constraints (about 20 kPa) indicate that the effect of mechanical impedance cannot be explained in terms of the increased external pressure causing a reduction in the rate of cell expansion (Wilson *et al.*, 1977; see also Feldman, 1984).

Perhaps the most interesting aspect of the root's response to mechanical impedance and soil pore size is its ability to modify the pattern of lateral branching. Goss (1977) showed that, in barley, 50 kPa of applied pressure shortened the region of initiation of laterals from a zone of about 30 mm (control) to a zone only 4 mm from the apex! In many cases the increased growth in the length of the lateral roots compensated for the reduction in length of the primary axes.

Goss & Russell (1980) further found that when a barley root was forced to grow around a glass bead, the site of lateral root initiation occurred preferentially on the outer (convex) side, whereas root hairs developed on the inner (concave) side of the root. This level of developmental complexity along with the examples cited above, clearly illustrate the high level of morphological and anatomical "plasticity" present within the cells of the root apex. This "plasticity" will be important during the life of the plant, but may be paramount in the early stages of germination, where the establishment of an adequate water supply is crucial.

Finally, given the obvious importance of these root characteristics, it is disappointing that at present "it is not possible to advance a detailed interpretation of the way root development is subsequently affected... Present uncertainty reflects the incompleteness of information on many aspects of the processes whereby growth is coordinated (Goss & Russell, 1980)."

Mucilage and the rhizosphere

According to Rougier (1981), Haberlandt originally proposed that root cap slime functions both as a lubricant and a protectant for the newly-forming root tip. These slime-, or mucilage-producing cells are found at the periphery of the root cap. In addition to their secretory functions, these cells are continually being sloughed off, so that the surface of the growing root becomes covered by a very heterogeneous material. This gelatinous material, which ensheaths roots growing in normal non-sterile soil, contains, in addition to soil particles, substances of both plant and microbial origin as well as living plant and microbial cells, and is generally termed mucigel (Rovira, Foster & Martin, 1979).

Over the ensuing seventy years since Haberlandt made his observations, considerable research effort has been expended in an attempt to elucidate the composition, properties and biological functions of this very interesting material. Several excellent reviews are available and the reader is referred to these for an extensive analysis of this literature (see Oades, 1978; Rovira *et al*. 1979; Rougier, 1981 and Northcote, 1982). I will briefly describe the site of synthesis of root cap slime and then discuss some of the more controversial functions attributed to the mucigel.

Although most research on slime production has been conducted on a few hydroponically grown plants, chiefly *Zea mays*, the details are generally applicable to most root systems. Cells produced by the root cap meristem develop into statocytes, containing numerous starch-filled amyloplasts (see next section for role played by these cells). Continued meristematic activity displaces these cells towards the root periphery and, at maturation, their cytoplasm undergoes marked changes (see Juniper, Gilchrist & Robins, 1977; Moore & McClelen, 1983). The starch reserves within the amyloplasts become degraded and the carbohydrate that they contained is processed via the endoplasmic reticulum and Golgi apparatus. Exocytosis of the secretory vesicles produced by the Golgi delivers the complex polysaccharides (or slime) into either the space between the cell wall and the plasmalemma, or into the wall proper, if the polysaccharide is small enough to pass through the interstices of the wall (Morré, Jones & Mollenhauer, 1976; Paull & Jones, 1975; Rougier, 1981).

The slime produced by various hydroponically-grown plants has been analyzed by several laboratories. However, as pointed out by Rougier (1981) and Chaboud (1983), the roots used were usually not grown under sterile conditions, so just what originated from the root cap cells was difficult to distinguish from metabolic products of the microbial flora. However, these analyses indicate that slime consists, in part, of a central β-1, 4-linked glucan that is rendered soluble by a coating of hydrophilic polysaccharides that are linked both covalently and non-covalently. The covalently-linked polysaccharides are thought to be relatively rich in galaturonic acid and fucose in regions near the central glucan (Wright & Northcote, 1976; Rougier, 1981).

The presence of fucose in this high molecular weight polysaccharide is a characteristic feature of the slime produced by the root cap. Vermeer & McCully (1981, 1982) used this feature to develop a lectin label that allowed them to follow the polysaccharide distribution in axenic (hydroponic) and field-grown maize plants. Other studies have been based on the detection of various chemical groups by more standard cytochemical methods (see examples cited in Rougier, 1981). Collectively, these studies reveal that the rhizosphere of the root is a more complex mucigel than previously envisaged. For example, Miki, Clarke & McCully (1980) found that in maize, wheat, barley, oats and sorghum the root surface is covered by two types of mucilage. A gelatinous material originates from the root cap and a firm, uniformly thick mucilage overlies the columnar epidermal cells. The epidermal mucilage has a thin outer and thicker inner layer distinct from the epidermal cell wall; periodic acid-Schiff staining indicated that the outer layer and the cell wall contain carboxyl groups which are absent from the thick inner layer. The root cap mucilage of these plants had an inner region with histochemical properties resembling those of the inner epidermal mucilage.

In a further study on field-grown maize plants, Vermeer & McCully (1982) observed that the rhizodermal sheath of the root is permeated by extracellular mucilage which is histochemically distinct from that present at the epidermal surface. This mucilage appears to originate from the root cap.

Finally, Foster (1981) conducted histochemical tests on root material of *Paspalum notatum* and *Triticum aestivum* that had been fixed, *in situ*, in the soil. He reported that he could distinguish, via various histochemical tests, between the polysaccharides secreted by the epidermis and those produced by soil fungi and bacteria. He also found that the nature of the mucilage and the boundaries between the cell wall and the soil depended on the age of the epidermal cells. In this context, a cuticle could be detected on the surface of the younger regions of the root (Foster, 1982). This level of chemical and spatial heterogeneity within the rhizosphere must surely reflect various roles of the mucigel in the physiological functioning of the root.

Numerous phenomenological observations have been reported that tend to support the hypothesis that mucigel production by the root tip (both root cap and epidermal cells) can respond interactively to changes within the immediate physical environment of the root. The production of mucigel is thought to be important in protecting younger parts of roots from desiccation during short periods of drought (see Greenland, 1979; Rougier, 1981). In an interesting study using fucose-specific lectins, Baldo, Reid & Boniface (1983) demonstrated that, in developing (maturing) wheat grains, a fucose-containing mucilage is already present around the root prior to germination. Such results support the claim that root cap mucilage functions as a protectant/lubricant.

The relationship between mucigel production and water stress is of particular interest, because crop plants are almost continuously exposed to varying levels of water stress. Recent work on axenic cultures of *Zea mays* (Chaboud, 1983) and *Oryza sativa* (rice)(Chaboud & Rougier, 1984) is important in this context. Their analyses of the root cap mucilage from these two root systems revealed a high content of galactose, arabinose and xylose. They suggested that the mucilage polymers could be related to, or could contain arabinogalactans; these compounds possess such properties as adhesiveness, water holding capacity and the ability to associate with other macromolecules (see Clarke, Anderson & Stone, 1979).

If a constituent of the mucigel does possess arabinogalactan-like properties this would certainly aid in the aggregation of soil particles around the root (see Oades, 1978). In addition, the strong water holding/binding properties would help to maintain the continuity of the water column from the soil to the root surface. Thus, under the development of water stress, the root may secrete more mucilage not only to facilitate its penetration into soil, but also to prevent its desiccation and to attempt to maintain the continuity of the contact between root and soil water.

Limited support is available in the literature for this role of the mucigel as an adaptation to water stress. Martin (1977) used an interesting experimental system to study the relationship between water stress and organic carbon release in soil-grown wheat roots. Individual wheat plants were grown in soil columns that had varying soil moisture content in the upper half, while the lower half of the column was saturated. By supplying $^{14}CO_2$ to the leaves, Martin could follow the release of organic carbon into the various regions of the soil. Although released organic carbon was reported as the summation of carbon secreted as mucilage and by cell lysis, the amount increased in zones of water stress, even though other sections of the root had adequate available water.

It has often been speculated that the carboxyl groups within the mucigel influence ion uptake into the root (Rougier, 1981; Chaboud & Rougier, 1984). In calcareous soils the high divalent cation binding affinity of the mucigel may protect the root tip from Ca^{2+} toxicity. In addition, the bound Ca^{2+} may form Ca^{2+}-galacturonic acid bridges between the different polysaccharide units, thereby modifying the physical characteristics of the mucigel (Cortez & Billes, 1982).

Aluminium toxicity in plants, which is often observed in acid soils, may be alleviated by the production of sufficient mucigel to protect the root meristem. In hydroponic experiments on *Vigna unguiculata* (cowpea) removal of the root cap mucliage resulted in an enhanced Al sensitivity of root elongation (Horst, Wagner & Marschner, 1982). Because mucilage has a high binding affinity for polyvalent cations (see Clarkson & Sanderson, 1969), Horst *et al.* (1982) suggested that mucilage may have an important ecological role in providing Al tolerance to plants growing in acid mineral soils. This hypothesis should be further investigated in field-based studies.

Attention has also been focused on the microbial utilization of the organic carbon released by cells of the root tip. Nitrogen fixing bacteria like *Azospirillum lipoferum* appear to be specifically attracted to certain plants, like maize (Baldani & Dobereiner, 1980), and these bacteria are associated with the mucigel (Umali-Garcia *et al*., 1980). However, whether these nitrogen fixing bacteria are important in terms of supplying organic nitrogen to the roots remains uncertain (see Beck & Gilmer, 1983; Whipp & Lynch, 1983). In any event, the quantity of the organic carbon released by the roots can be a large proportion of the total amount translocated into the tip (see chapter 6 for more discussion of this point). Thus, this area should remain active in future research. The statement made by Chaboud & Rougier (1981) that "a greater understanding of the nature of the mucigel, i.e., of its diverse components, would be of use in analyzing potential extracellular recognition phenomena of both benevolent and pathogenic microorganisms", seems to identify an important area for future studies. Elucidation of the mechanism(s) by which the plant root system may recognize and respond to pathogenic attack could be of value in developing a complete picture of the way the root system responds to stress phenomena in general.

GRAVITROPISM IN ROOTS

General background

Gravity is an important component of the root's physical environment and in general a positive response is observed at the root tip. Excellent reviews of plant gravitropism have been provided by Audus (1975), Jackson and Barlow (1981), Juniper (1976), Volkmann & Sievers (1979), and Wilkins (1984). We shall examine the gravitropic phenomenon in detail because it represents an excellent system in which to study regulation in a complex organ, the root. And of course, these regulatory events involve modulation of physiological processes occurring in the cells of the root apex.

At the beginning of the 1980s there appeared to be a consensus that roots responded to gravity in the following manner. The statenchyma, situated within the centre of the root cap, functions as the site of graviperception. In these cells specialized plastids, called amyloplasts, contain large starch granules. Because of their greater density in comparison with the surrounding cytosol, the amyloplasts, the presumed statoliths, sediment under the influence of gravity. In a vertically oriented root, the amyloplasts rest near the lower surface of the cell; i.e., the end of the cell nearest to the actual tip of the root cap, and the rate of cell elongation within the growth zone of the root conformed to circumferential symmetry. However, displacement of the root away from this orientation results in a redistribution of the amyloplasts. Although the mechanism of transduction is uncertain, this movement of the amyloplasts was thought to result in the release of an inhibitory plant growth substance (hormone, IAA, ABA or?) from the root cap. An asymmetric distribution of this putative substance, in the region of cell elongation (i.e., more in the lower than in the upper half of the root), would elicit root curvature. This is a summary of the Cholodny-Went hypothesis (see Went & Thimann, 1937). Unfortunately, to demonstrate the phenomenon of root gravitropism has been easier than to test the prediction of the Cholodny-Went hypothesis.

Hormone involvement: ABA, IAA or ?

Jackson & Barlow (1981) provided an excellent analysis of the literature on the role of plant growth regulators in root gravitropism. Since their review several important findings have been published. Pilet & Ney (1981) have now answered one of the questions raised by Jackson & Barlow. As illustrated in Fig. 4, curvature of the maize root, in response to gravity, is obtained by an initial marked decrease in growth rate on the lower side and a slight stimulation in growth rate on the upper side. However, the mechanism responsible for this modulation in growth remains controversial. That a concentration gradient of the hormone indole-3-acetic acid (IAA) can elicit a stimulation of growth at low concentrations and an inhibition at slightly higher levels, is clearly illustrated by experiments with segments of maize roots (Pilet, Elliott & Moloney, 1979). The important question is whether these experiments with excised segments of root can be extrapolated to the intact root system.

Using a very sensitive enzyme immunoassay and radioimmunoassay for IAA and abscisic acid (ABA), Mertens & Weiler (1983) studied the distribution of these two growth regulators in gravistimulated maize roots. Although they could detect endogenous IAA and ABA, they could not detect the predicted increased concentrations on the lower side of the gravitropic root. This result suggests that either these hormones are not involved in the gravitropic response, or the technique used does not differentiate between total hormone present and its actual distribution within active and nonactive pools.

It is important to point out that the immunoassay results are in contradiction to the results of Pilet & Rivier (1981), who used gas chromatography-mass spectroscopy to analyze for endogenous ABA levels. However, as will be discussed shortly, present consensus is that ABA may not be directly involved in gravitropism.

The recent work of Lee, Mulkey & Evans (1984) also implicates IAA in the graviresponse mechanism. Treatment of maize or pea roots with inhibitors of auxin transport (morphactin, naphthylphthalamic acid, and

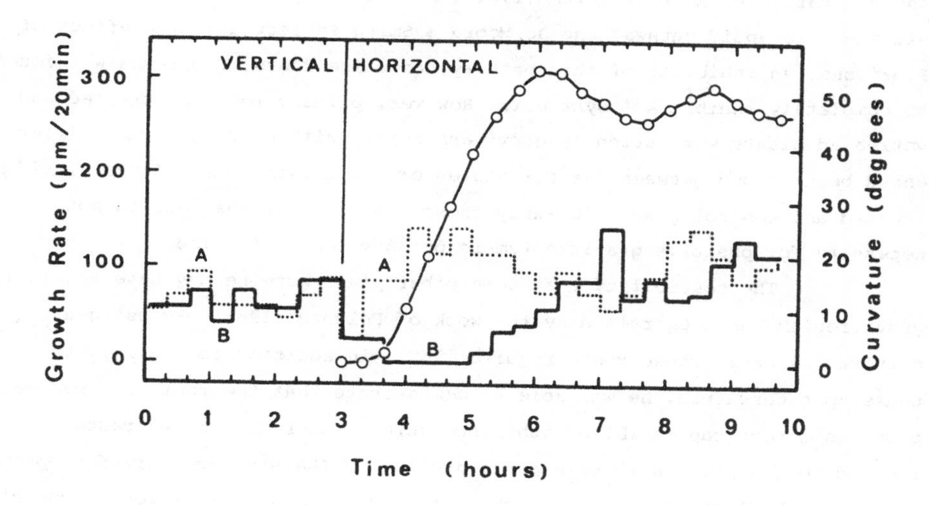

Figure 4. Growth rate of the two sides of a primary maize root that was first kept for 3 hours in the vertical position, and then placed horizontally for 7 hours. The gravitropic response (downward curvature) is plotted in degrees of deflection from the horizontal (O). A, upper side, B, lower side of maize root (Redrawn from Pilet & Ney, 1981).

2,3,5-triodobenzoic acid) prevented the gravitropic response. This is consistent with their earlier finding, in that these auxin transport inhibitors also suppressed the asymmetric acid efflux that appears to be a characteristic of gravistimulated maize roots (Mulkey & Evans, 1982). However, in their recent work, Lee *et al.* (1984) reported that these auxin transport inhibitors also prevented the gravity-induced polar movement of Ca^{2+} across the *root tip*. Thus, the question of IAA involvement remains to be resolved.

Jackson & Barlow (1981) presented a thorough examination of the experimental results concerning the role of ABA in gravitropism, and they concluded that "the evidence that ABA is the inhibitor seems poor at present." Several recent studies support their conclusion (see, for example, Mulkey, Evans & Kuzmanoff, 1983). Perhaps the most definitive results were those obtained by Moore & Smith (1984). ABA is thought to be synthesized via the carotenoid pathway and so Moore & Smith investigated the effect of Fluridone, an inhibitor of this pathway, on maize roots. Fluridone was found to completely inhibit ABA synthesis. However, primary roots of treated and untreated plants were strongly graviresponsive, with no significant differences being found between the curvatures or the growth rates of the Fluridone-treated and control plants. Clearly these results show that ABA is not necessary for positive gravitropism in primary roots of maize.

The possibility that some other plant hormone may have a role in gravitropism has been raised by the work of Feldman (1981). By using a cultivar of maize whose roots require light, in addition to gravity, to cause root curvature, he was able to demonstrate that the root cap, *per se*, produces a root cap inhibitor substance which causes dark-grown roots to respond to gravity. In these experiments 10^{-9} M IAA was necessary for optimal root cap inhibitor (RCI) production, but *in vivo* this hormone would probably be supplied from its site of synthesis in the meristematic zone of the apex (Feldman, 1980).

In subsequent studies on this light-dependent maize cultivar, Feldman (1983) and Feldman & Gildow (1984) showed that light stimulates protein synthesis (but not DNA synthesis) that is necessary for the gravitropic response of the maize root. Using electrophoretic techniques, they were able to single out specific proteins whose levels within the root cap were influenced by light. Interestingly, several of these proteins, for which synthesis was stimulated by light, appeared to turn over rapidly. It will be very interesting to follow the developments on this RCI substance.

Role of Ca^{2+} in root gravitropism

Although Ca^{2+} has long been known to be important in polar auxin transport (see de la Fuente & Leopold, 1973), more recent studies indicate that this divalent cation may be involved in the early stages of transduction of the regulatory signal(s), from the root cap, to the zone of elongation, where curvature occurs.

If the statoliths (amyloplasts), in the statocytes of the root cap, are involved in the initial perception of a change in the gravitational field, the first transduction step must occur within these cells. At present no consensus exists as to the nature of this signal, but a change in cytosolic Ca^{2+} may be involved. For a long time it has been proposed that the amyloplasts may form some special contact with the endoplasmic reticulum (ER) in the statenchyma (see Sievers & Volkmann, 1972; Hensel, 1984; and also Barlow, Hawes & Horne, 1984; and Olsen *et al.*, 1984, for a differing viewpoint). The physical change in location of the amyloplasts, in response to gravity, may elicit a Ca^{2+} response from either the amyloplasts, or the ER. Chandra *et al.* (1982) used an ion microscope to show that Ca^{2+} is present in the root cap statoliths of maize, pea and lettuce. In addition, Buckhout, Heyder-Caspers & Sievers (1982) and Buckhout (1983) have developed a technique that allows them to isolate the ER from the root cap region. They found that the vesicles in this putative ER fraction took up Ca^{2+} via an ATP-dependent process that appeared to be a direct CA^{2+}-translocating ATPase. There are also several reports for plant tissues, other than root cap cells, which indicate that a Ca^{2+} efflux system is probably present in the plasmalemma (see Dieter & Marme, 1983; Dieter, 1984).

Since Biro *et al.* (1982) found that chlorpromazine, an inhibitor of the Ca^{2+}-activated form of calmodulin, significantly inhibited the negative gravitropic response of the oat coleoptile, it seems plausible to suggest that a calmodulin-mediated signal(s) may well be involved in the earliest stages of graviperception. We know of no reports on the effects of calmodulin inhibitors on the gravitropic response in roots. However, Buckhout (1984) found that calmodulin seems not to be involved in regulating Ca^{2+} transport across ER membrane vesicles prepared from *Lepidium sativum* roots. This area of research should be pursued to test the Ca^{2+}-calmodulin hypothesis.

An alternative Ca^{2+} signalling system in animal tissues, involving Inositol 1,4,5-triphosphate ($InsP_3$), has recently received considerable attention (see Nishizuka, 1984). In this system, hormone binding to a receptor site in the plasma membrane is presumed to activate an inositol diesterase and this eventually leads to the production of $InsP_3$ (Berridge, 1983). Prentki *et al.* (1984) showed that this $InsP_3$ causes the rapid release of Ca^{2+} from a microsomal fraction of rat insulinoma but not from mitochondria, and they propose that $InsP_3$ functions as a cellular messenger inducing the mobilization of Ca^{2+} from the endoplasmic reticulum. An interesting study on $InsP_3$, which illustrates a way in which this system may function in the statocytes, is that conducted on ventral photo-receptors. A single photon can cause approximately 1,000 ionic channels to open in the receptor of the eye. This must require an internal transmitter and Fein *et al.* (1984) demonstrated that injection of $InsP_3$ excited and adapted ventral photoreceptors in a manner similar to light.

On a speculative note, sedimenting amyloplasts may cause a perturbation in the microtubular arrangement in the statocytes and so move plasma membrane-bound, or associated, receptors that activate the $InsP_3$ system. This $InsP_3$ may also be involved in opening ionic (Ca^{2+}?) channels in plant cells or it may mediate in the ER release of Ca^{2+}.

Evans and his coworkers have shown that the gravitropic response in roots of maize and pea can be modified by adjusting the exogenous Ca^{2+} level. Lee, Mulkey & Evans (1983a) found that application of Ca^{2+} chelating agents, like EDTA or EGTA (held in small agar blocks), to the tips of maize roots caused a loss of gravitropic sensitivity. Non-specific effects cannot be discounted, because an extremely high concentration (50 mM) was used. But, since replacing the chelating agent with $CaCl_2$ restored the root's gravitropic sensitivity, the results are probably valid. Placement of agar blocks containing $CaCl_2$ on one side of the root cap of a vertically oriented root caused curvature towards the Ca^{2+}, whereas application of a block containing EDTA caused curvature away from the EDTA or Ca^{2+}-diminished side of the root. However, application of $CaCl_2$ to the tips of decapped maize roots, which in this state are insensitive to gravity, did not restore this ability. Thus, Ca^{2+}, *per se*, cannot be a direct signal, but seems rather to be one of the factors necessary for the transduction of the growth regulating signal.

Further work by this group illustrated that $^{45}Ca^{2+}$ movement across the root tip depends strongly on the position of the root. When oriented horizontally the maize root cap showed a strong asymmetry to the movement of $^{45}Ca^{2+}$; movement from top to bottom increased, while that from the bottom to the top decreased (Lee, Mulkey & Evans, 1983b). Similar results were obtained on the graviresponsive roots of pea, but no such asymmetry was detected on onion roots which do not respond to gravity (Lee *et al.*, 1984). A further finding was that pretreatment of maize and pea roots with auxin transport inhibitors prevented both the gravitropic response and the polar movement of $^{45}Ca^{2+}$ across the root cap. These results suggest that in some way Ca^{2+} plays a key role in linking gravistimulation to the gravitropic growth response in roots. Lee *et al.* (1984) suggest that the increased level of Ca^{2+} on the lower side of the root cap "may lead to enhanced acropetal movement of auxin into the elongation zone on the lower side of the root", i.e., they still favour the Cholodny-Went hypothesis.

Geoelectric phenomena in roots

The role of electric currents has recently been invoked in terms of mediating the graviperceptive response of roots. We will examine both the old concept of the geoelectric response of roots and the recent electrical data to ascertain whether a biophysical mechanism is involved in the transduction phase of graviperception.

The original geoelectric theory was based on the early observations of Bose (1907) and Brauner (1927), who found that coleoptiles and roots, when placed in the horizontal position, developed an electric polarity across the organ. In roots, the lower side was positively charged. Thus since auxin is a weak acid, it would be present as an anion at pH 5 to 6, and so the then presumed growth-active hormone would be attracted to the positively charged lower surface where it would inhibit cell elongation.

A difficulty in these early electrical studies was to make good electrical contacts with the tissue. The problem was solved by Grahm & Hertz (1962) and Grahm (1964), who developed an ingenious vibrating-reed electrode that could detect the surface electric field of gravistimulated coleoptiles without making contact with the surface of the organ. Using this technique they were able to show that indeed an asymmetric electric potential is established following gravistimulation, however, the time course was much slower than the almost immediate response reported by Brauner (1927). Grahm (1964) clearly showed that the electrical phenomenon had a lag phase of 15 minutes which was almost exactly the same as the observed latent period for curvature response in the coleoptile. In view of this, Grahm concluded that "the geoelectric effect is a secondary phenomenon of asymmetric auxin distribution", and thus cannot be the factor responsible for auxin movement.

The concept of electrical events being involved in gravistimulation in roots received no further attention until the work of Behrens, Weisenseel & Sievers (1982). These workers used the *Lepidium sativum* root system and applied the external vibrating electrode technique to study the influence of gravistimulation on the electric fields that are normally present around roots (see Scott, 1967). Although their experimental system was not set up for making optimal measurements, Behrens *et al.* (1982) found that the electric field, in the region of the root tip, underwent a rapid change (within 30 seconds), in response to a change in the gravitational field (see Fig. 5). Gravitropic curvature was first visible approximately 10 minutes after placing the root in the horizontal position. Thus, since the change in current pattern in the root cap region precedes root curvature, Behrens *et al.* (1982) suggested that the current may be involved in the transduction of information from the root cap to the elongation zone, following graviperception in the cap.

Sievers *et al.* (1984) and Behrens, Gradmann & Sievers (1985) pursued this concept further by making electrophysiological measurements on the cells of the *Lepidium* root cap. Their findings were indeed very interesting. As illustrated in Fig. 6, the membrane potential of statocytes on the upper and lower surface of the gravistimulated root responded in an asymmetric manner. In statocyte cells on the lower surface the membrane potential was transiently depolarized, while cells on the upper surface responded by developing a more negative potential across the plasmalemma. Sievers *et al.* (1984) developed a model to relate cytological changes, induced by a change in gravistimulation, to extracellular current and membrane potential (Fig. 7).

As discussed in an earlier section (4.3), there are numerous ways in which the initial events of perception could be transduced into a membrane depolarizing signal (Ca^{2+}-calmodulin or Ca^{2+}-$InsP_3$ may be involved). But we do not know what ionic fluxes are responsible for the transient depolarization or what new transmembrane ion flux pattern is responsible for the change in extracellular current at the root cap (see Fig. 5). Finally, if these electrical changes are part of the early perception-transduction phase of graviperception (and the data are certainly compelling), how do these current changes elicit the necessary modulation in growth rates in the far removed zone of elongation?

CONCLUSIONS

Roots, functioning not in laboratory culture but in the soil, must be capable of making a wide range of adjustments in response to their changing physical environment. Many of the examples cited to illustrate both the functional aspects of the cells of the root apex, and how these cells may respond to environmental changes, have been drawn from laboratory-

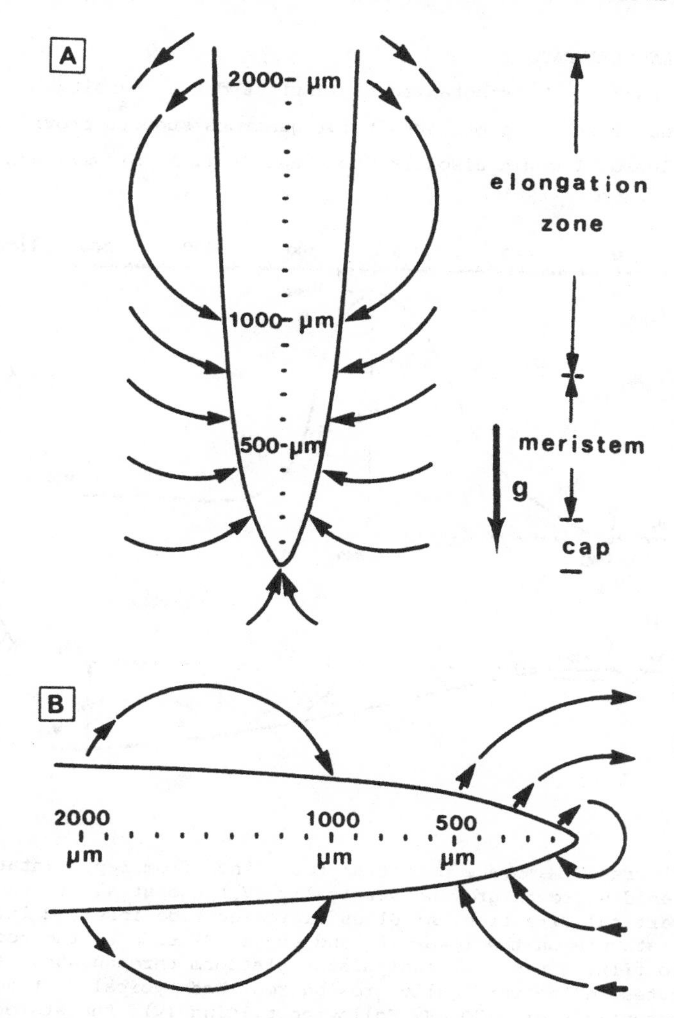

Figure 5. Current pattern associated with the root tip of *Lepidium*. A: Symmetrical pattern during vertical growth. B: Asymmetric pattern established in the root cap 3 minutes after placing root in the horizontal position. (By courtesy of H.M. Behrens.)

based studies. This is a consequence of the difficulty of designing experiments on soil-grown root systems when so much basic information on the functional aspects of root cells is still lacking. When such knowledge becomes available and soil-grown plants become widely used, we will gain greater appreciation of the biochemical/biophysical control systems that must function within the root to allow it to survive.

ACKNOWLEGMENTS

I thank the Forstbotanisches Institut der Universitat Gottingen, and the Deutsche Forschungsgemeinschaft for generous support provided during my sabbatical leave. I would also like to thank Prof. A. Sievers and his

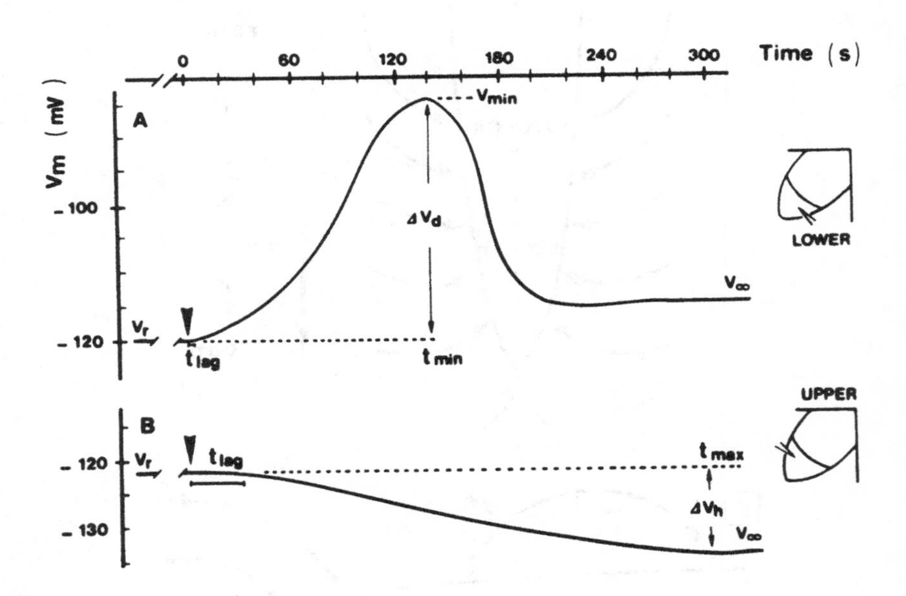

Figure 6. Membrane potential recordings from representative *Lepidium* roots growing vertically (V_r) and at 45° to the vertical direction. A: Glass microelectrode inserted into statocyte on the lower (A) and upper (B) side of the root prior to tilting the micromanipulator platform through 45°. Statocytes in the vertically growing root had typical resting potentials of -120 mV. Following tilting (V), the statocytes on the lower side of the root transiently depolarized to approx. -84 mV (V_{min}) and then repolarized back to a value of approx. -115 mV (V_∞). The statocytes on the upper side of the root hyperpolarized their membrane potential, to a new table value of approx. -134 mV, upon tilting. (From Sievers *et al.*, 1984).

colleagues for providing their manuscripts prior to publication. This work would not have been possible without the "long-range" support of my laboratory in Davis, and especially the efforts of Clyde Wilson and Dr. Leon Kochian. Finally I would like to thank Frau Wehr and Lyn Noah for their assistance in preparing the manuscript.

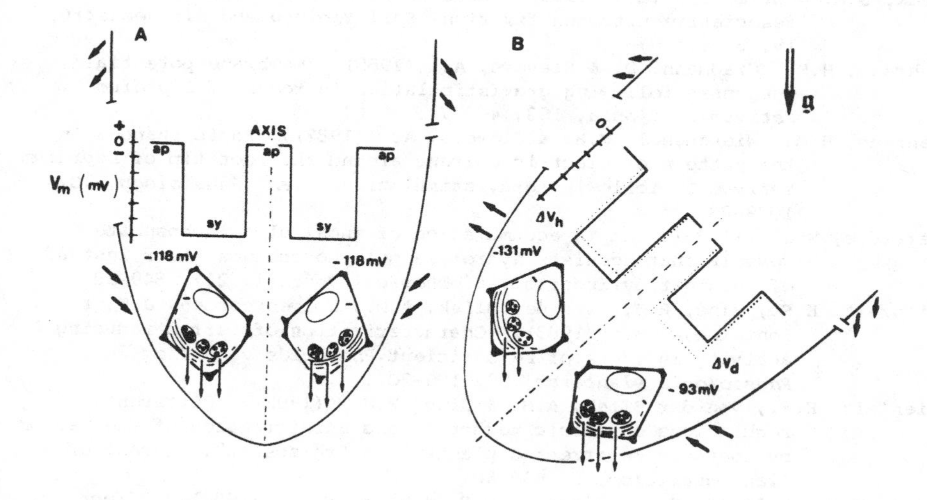

Figure 7. Synopsis of early graviresponses in *Lepidium* root tips. A: In vertically growing roots, three symmetrical patterns are observed; sedimentation of the amyloplasts on the ER complex (thin arrows in the g direction), the potential profiles through the symplast (sy) and apoplast (ap) and the direction of external current flow (thick arrows). B: Following tilting at 45° from the vertical direction, these patterns become asymmetric; the ER complex is differentially compressed by the sedimenting amyloplasts (thin arrows) and a potential depolarization (ΔV_d) in the physically lower and hyperpolarization (ΔV_h) in the upper statocytes. These changes generate a net asymmetrical current pattern across the root cap, the site of graviperception (From Sievers *et al.*, 1984).

REFERENCES

Ambler, J.E., Brown, J.C. & Gauch, H.G. (1971). Sites of iron reduction on soybean plants. *Agronomy Journal*, 63, 95-7.

Audus, L.J. (1975). Geotropism in roots. In *The Development and Function of Roots*, ed. J.G. Torrey & D.T. Clarkson, pp.327-63. London: Academic Press.

Baldani, V.L. & Dobereiner, D.J. (1980). Host-plant specificity in the infection of cereals with *Azospirillum* spp. *Soil Biology and Biochemistry*, 12, 443-9.

Baldo, B.A., Reid, A.L. & Boniface, P.A. (1983). Lectins as cytochemical probes of the developing wheat grain. IV. Demonstration of mucilage containing L-fucose associated with roots in ungerminated grain. *Australian Journal of Plant Physiology*, 10, 459-70.

Barlow, P.W., Hawes, C.R. & Horne, J.C. (1984). Structure of amyloplasts and endoplasmic reticulum in the root caps of *Lepidium sativum* and *Zea mays* observed after selective membrane staining and by high-voltage electron microscopy. *Planta*, 160, 363-71.

Beck, S.M. & Gilmour, C.M. (1983). Role of wheat root exudates in associative nitrogen fixation. *Soil Biology and Biochemistry*, 15, 33-8.

Behrens, H.M., Gradmann, D. & Sievers, A. (1985). Membrane-potential responses following gravistimulation in roots of *Lepidium sativum* L. *Planta*, 163, 463-72.

Behrens, H.M., Wiesenseel, M.H. & Sievers, A. (1982). Rapid changes in the pattern of electric current around the root tip of *Lepidium sativum* L. following gravistimulation. *Plant Physiology*, 70, 1079-83.

Berridge, M.J. (1983). Rapid accumulation of inositol trisphosphate reveals that agonists hydrolyse polyphosphoinositides instead of phosphatidylinositol. *Biochemical Journal*, 212, 849-58.

Bienfait, H.F., Bino, R.J., van der Bliek, A.M., Duivenvoorden, J.F. & Fontaine, J.M. (1983). Characterization of ferric reducing activity in roots of Fe-deficient *Phaseolus vulgaris*. *Physiologia Plantarum*, 59, 196-202.

Bienfait, H.F., van der Bliek, A.M. & Bino, R.J. (1982). Different regulations on ferric reduction and acidification of the medium by roots of Fe-stressed plants in a "rhizostat". *Journal of Plant Nutrition*, 5, 447-50.

Biro, R.L., Hale II, C.C., Wiegand, O.F. & Roux, S.J. (1982). Effects of chlorpromazine on gravitropism in *Avena* coleoptiles. *Annals of Botany*, 50, 737-45.

Bose, J.C. (1907). *Comparative Electrophysiology*. London: Longmans and Green.

Brauner, L. (1927). Untersuchungen uber das geoelektrische Phanomen. I. *Jahrbucher fur wissenschaftliche Botanik*, 66, 381-428.

Brown, J.C. & Ambler, J.E. (1974). Iron-stress response in tomato (*Lycopersicon esculentum*). I. Sites of iron reduction, absorption and transport. *Physiologia Plantarum*, 31, 221-4.

Buckhout, T.J. (1983). ATP-dependent Ca^{2+} transport in endoplasmic reticulum isolated from roots of *Lepidium sativum* L. *Planta*, 159, 84-90.

Buckhout, T.J. (1984). Characterization of Ca^{2+} transport in endoplasmic reticulum membrane vesicles from *Lepidium sativum* L. roots. *Plant Physiology*, 76, 962-7.

Buckout, T.J., Heyder-Caspers, L. and Sievers, A. (1982). Fractionation and characterization of cellular membranes from root tips of garden cress (*Lepidium sativum* L.). *Planta*, 156, 108-16.
Chaboud, A. (1983). Isolation, purification and chemical composition of maize root cap slime. *Plant and Soil*, 73, 395-402.
Chaboud, A. and Rougier, M. (1981). Secretions racinarier mucilagineuses et role dans la rhizosphere. *Annals of Biology*, 20, 313-26.
Chaboud, A. and Rougier, M. (1984). Identification and Localization of sugar components of rice (*Oryza sativa* L.) root cap mucilage. *Journal of Plant Physiology*, 116, 323-30.
Chandra, S., Chabot, J.F., Morrison, G.H. and Leopold, A.C. (1982). Localization of calcium in amyloplasts of root-cap cells using ion microscopy. *Science*, 216, 1221-3.
Clarke, A.E., Anderson, R.L. & Stone, B.A. (1979). Form and function of arabinogalactans and arabinogalactan-proteins. *Phytochemistry*, 18, 521-40.
Clarkson, D.T. and Sanderson, J. (1969). The uptake of a polyvalent cation and its distribution in the root apices of *Allium cepa*: Tracer and autoradiographic studies. *Planta*, 89, 136-54.
Clarkson, D.T. and Sanderson, J. (1978). Sites of absorption and translocation of iron in barley roots. *Plant Physiology*, 61, 731-6.
Cortez, J. and Billes, G. (1982). Role of calcium ions in formation of mucilage in the root cap of *Zea mays*. *Acta Oecologica* Serie 3, *Oecologia Plantarum*, 3, 67-78.
dela Fuente, R.K. and Leopold, A.C. (1973). A role for calcium in auxin transport. *Plant Physiology*, 51, 845-7.
Dieter, P. (1984). Calmodulin and calmodulin-mediated processes in plants. *Plant, Cell and Environment*, 7, 371-80.
Dieter, P. & Marme, D. (1983). The effect of calmodulin and far-red light on the kinetic properties of the mitochondrial and microsomal calcium-ion transport system for corn. *Planta*, 159, 277-81.
Fein, A., Payne, R., Corson, D.W., Berridge, M.J. and Irvine, R.F. (1984). Photoreceptor excitation and adaptation by inositol 1,4,5-trisphosphate. *Nature*, 311, 157-60.
Feldman, L.J. (1980). Auxin biosynthesis and metabolism in isolated roots of *Zea mays*. *Physiologia Plantarum*, 49, 145-50.
Feldman, L.J. (1981). Root cap inhibitor formation in isolated root caps of *Zea mays*. *Journal of Experimental Botany*, 323, 779-88.
Feldman, L.J. (1983). Light-enhanced protein synthesis in gravitropically stimulated root caps of corn. *Plant Physiology*, 72, 833-6.
Feldman, L.J. (1984). Regulation of root development. *Annual Review of Plant Physiology*, 35, 223-42.
Feldman, L.J. & Gildow, V. (1984). Effects of light on protein patterns in gravitropically stimulated root caps of corn. *Plant Physiology*, 74, 208-12.
Foster, R.C. (1981). The ultrastructure and histochemistry of the rhizosphere. *New Phytologist*, 89, 263-73.
Foster, R.C. (1982). The fine structure of epidermal cell mucilages of roots. *New Phytologist*, 91, 727-40.
Goss, M.J. (1977). Effects of mechanical impedance on root growth in barley (*Hordeum vulgare* L.). I. Effects on the elongation and branching of seminal root axes. *Journal of Experimental Botany*, 31, 577-88.
Grahm, L. (1964). Measurements of geoelectric and auxin-induced potentials in coleoptiles with a refined vibrating reed electrode technique. *Physiologia Plantarum*, 17, 231-61.

Grahm, L. & Herz, C.H. (1962). Measurements of the geoelectric effect in coleoptiles by a new technique. *Physiologia Plantarum*, 15, 96-114.

Greenland, D.J. (1979). The physics and chemistry of the soil-root interface: some comments. In *The Soil-Root Interface*, ed. J.L. Harley & R.S. Russell, pp. 83-98. London: Academic Press.

Gunning, B.E. (1977). Transfer cells and their roles in transport of solutes in plants. *Science Progress, Oxford*, 64, 539-68.

Hensel, W. (1984). A role of microtubules in the polarity of statocytes from roots of *Lepidium sativum* L. *Planta*, 162, 404-14.

Horst, W.J., Wagner, A. & Marschner, H. (1982). Mucilage protects root meristem from aluminium injury. *Zeitschrift für Pflanzenphysiologie*, 105, 435-44.

Jackson, M.B. & Barlow, P.W. (1981). Root gravitropism and the role of growth regulators from the cap: a re-examination. *Plant, Cell and Environment*, 4, 107-23.

Juniper, B.E. (1976). Geotropism. *Annual Review of Plant Physiology*, 27, 385-406.

Juniper, B.E., Gilchrist, A.J. & Robins, R.J. (1977). Some features of secretory systems in plants. *Histochemistry Journal*, 9, 659-80.

Kramer, D. (1983). Genetically determined adaptations in roots to nutritional stress: correlations of structure and function. *Plant and Soil*, 72, 167-73.

Lee, J.S., Mulkey, T.J. & Evans, M.L. (1983a). Reversible loss of gravitropic sensitivity in maize roots after tip application of calcium chelators. *Science*, 220, 1375-6.

Lee, J.S., Mulkey, T.J. & Evans, M.L. (1983b). Gravity-induced polar transport of calcium across root tips of maize. *Plant Physiology*, 73, 874-6.

Lee, J.S., Mulkey, T.J. & Evans, M.L. (1984). Inhibition of polar calcium movement and gravitropism in roots treated with auxin-transport inhibitors. *Planta*, 160, 536-43.

Marschner, H., Romheld, V. & Ossenberg-Neuhaus, H. (1982). Rapid method for measuring changes in pH and reducing processes along roots of intact plants. *Zeitschrift fur Pflanzenphysiologie*, 105, 407-16.

Martin, J.K. (1977). Effect of soil moisture on the release of organic carbon from wheat roots. *Soil Biology and Biochemistry*, 9, 303-4.

Mertens, R. & Weiler, E.W. (1983). Kinetic studies on the distribution of endogenous growth regulators in gravireacting plant organs. *Planta*, 158, 339-48.

Miki, N.K., Clarke, K.J. & McCully, M.E. (1980). A histological and histochemical comparison of the mucilages on the root tips of several grasses. *Canadian Journal of Botany*, 58, 2581-93.

Moore, R. & McClelen, C.E. (1983). Ultrastructural aspects of cellular differentiation in the root cap of *Zea mays*. *Canadian Journal of Botany*, 61, 1566-72.

Moore, R. & Smith, J.D. (1984). Growth, graviresponsiveness and abscisic-acid content of *Zea mays* seedlings treated with Fluridone. *Planta*, 162, 342-4.

Morré, D.J., Jones, D.D. & Mollenhauer, H.H. (1967). Golgi apparatus mediated polysaccharide secretion by outer root cap cells of *Zea mays*. I. Kinetics and secretory pathway. *Planta*, 74, 386-301.

Mulkey, T.J. & Evans, M.L. (1982). Suppression of asymmetric acid efflux and gravitropism in maize roots treated with auxin transport inhibitors or sodium orthovanadate. *Journal of Plant Growth Regulators*, 1, 259-65.

Mulkey, T.J., Evans, M.L. & Kuzmanoff, K.M. (1983). The kinetics of abscisic acid action on root growth and gravitropism. *Planta*, 157, 150-7.

Northcote, D.H. (1982). The synthesis and transport of some plant glycoproteins. *Philosophical Transactions of the Royal Society, London*, Series B, 300, 195-206.

Nishizuka, Y. (1984). The role of protein kinase C in cell surface signal transduction and tumor promotion. *Nature*, 308, 693-8.

Oades, J.M. (1978). Mucilages at the root surface. *Journal of Soil Science*, 29, 1-16.

Olsen, G.M., Mirze, J.I., Maher, E.P. & Iversen, T-H. (1984). Ultrastructure and movements of cell organelles in the root cap of agravitropic mutants and normal seedlings of *Arabidopsis thaliana*. *Physiologia Plantarum*, 60, 523-31.

Paull, R.E. & Jones, R.L. (1975). Studies on the secretion of maize root-cap slime. II. Localization of slime production. *Plant Physiology*, 56, 307-12.

Pilet, P.E., Elliott, M.C. & Moloney, M.M. (1979). Endogenous and exogenous auxin in the control of root growth. *Planta*, 146, 405-8.

Pilet, P.E. & Rivier, L. (1981). Abscisic acid distribution in horizontal maize root segments. *Planta*, 153, 453-8.

Prentki, M., Biden, T.J., Janjie, D., Irvine, R.F., Berridge, M.J. & Wollheim, C.B. (1984). Rapid mobilization of Ca^{2+} from rat insulinoma microsomes by inositol-1,4,5-triphosphate. *Nature*, 309, 562-4.

Römheld, V. & Kramer, D. (1983). Relationship between proton efflux and rhizodermal transfer cells induced by iron deficiency. *Zeitschrift fur Pflanzenphysiologie*, 113, 73-83.

Römheld, V. & Marschner, H. (1981a). Rhythmic iron stress reáctions in sunflower at suboptimal iron supply. *Physiologia Plan* 53, 347-53.

Römheld, V. & Marschner, H. (1981b). Iron deficiency stress induced morphological and physiological changes in root tips of sunflower. *Physiologia Plantarum*, 53, 354-60.

Rougier, M. (1981). Secretory activity of the root cap. In *Encyclopedia of Plant Physiology*, New Series, Volume 13 B, Plant Carbohydrates II, ed. W. Tanner and F.A. Leowus, pp. 542-74. Berlin: Springer-Verlag.

Rovira, A.D., Foster, R.C. & Martin, J.K. (1979). Origin, nature and nomenclature of the organic materials in the rhizosphere. In *The Soil-Root interface*, ed. J.L. Harley and R.S. Russell, pp. 1-4. London: Academic Press.

Scott, B.I.H. (1967). Electric fields in plants. *Annual Review of Plant Physiology*, 18, 409-18.

Sievers, A., Behrens, H.M., Buckhout, T.J. & Gradmann, D. (1984). Can a Ca^{2+} pump in the endoplasmic reticulum of the *Lepidium* root be the trigger for rapid changes in membrane potential after gravistimulation? *Zeitschrift fur Pflanzenphysiologie*, 114, 195-200.

Sievers, A. & Volkmann, D. (1972). Verursacht differentieller Druck der Amyloplasten auf ein komplexes Endomembran-system die Geoperzeption in Wurzeln? *Planta*, 102, 160-72.

Sijmons, P.C. & Bienfait, H.F. (1983). Source of electrons for extracellular Fe(III) reduction in iron-deficient bean roots. *Physiologia Plantarum*, 59, 409-15.

Torii, K. & Laties, G.G. (1966). Dual mechanisms of ion uptake in relation to vacuolation in corn roots. *Plant Physiology*, 41, 863-70.

Umali-Garcia, M., Hubbell, D.H., Gaskins, M.H. & Dazzo, F.B. (1980). Association of *Azospirillum* with grass roots. *Applied Environmental Microbiology*, 39, 219-26.

Vermeer, J. & McCully, M.E. (1981). Fucose in the surface deposits of axenic and field grown roots of *Zea mays* L. *Protoplasma*, 109, 233-48.

Vermeer, J. & McCully, M.E. (1982). The rhizosphere in *Zea*: new insight into its structure and development. *Planta*, 156, 45-61.

Volkmann, D. & Sievers, A. (1979). Graviperception in multicellular organisms. In *Encyclopedia of Plant Physiology*, New Series, Volume 7, Physiology of Movements, ed. W. Haupt and M.E. Feinleib, pp. 573-600. Berlin: Springer-Verlag.

Went, F.W. & Thimann, K.V. (1937). *Phytohormones*. London: Macmillan.

Whipps, J.M. & Lynch, J.M. (1983). Substrate flow and utilization in the rhizosphere of cerials. *New Phytologist*, 95, 605-23.

Wilkins, M.B. (1984). Gravitropism. In *Advanced Plant Physiology*, ed. M.B. Wilkins, pp. 163-200. London: Pitman.

Wilson, A.J., Robards, A.W. & Goss, M.J. (1977). Effects of mechanical impedance on root growth in barley, *Hordeum vulgare* L. II. Effects on cell development in seminal roots. *Journal of Experimental Botany*, 28, 1216-27.

Wright, K. & Northcote, D.H. (1976). Identification of β-1,4-glucan chains as part of a fraction of slime synthesized within the dictyosomes of maize root caps. *Protoplasma*, 88, 225-39.

SELECTED ASPECTS OF THE STRUCTURE AND DEVELOPMENT OF FIELD-GROWN ROOTS WITH SPECIAL REFERENCE TO MAIZE

M.E. McCully

INTRODUCTION

The understanding of root function depends upon the interpretation of physiological data in relation to structural features. But anatomical data as available in standard treatises of plant anatomy is, for herbaceous roots at least, derived from studies of the primary roots of young, laboratory-grown seedlings. This standard anatomy does not cover developmental changes which may occur beyond a few cm proximal to a growing root tip. Nor does it give a clue to the heterogeneity of roots within a root system, nor describe features unique to field-grown roots. There has been some failure to appreciate the inadequacy of such standard information when it is applied to situations other than those pertaining to laboratory-grown seedling root tips.

There are, of course, descriptions of root structure *in situ* in the soil, particularly with reference to the root-soil interface (e.g. Foster, Rovira & Cock, 1983, and references therein). The complexity of the features of the root systems seen in such micrographs emphasizes the inadequacy of the standard anatomy to cover field situations. In general, however, these studies of rhizosphere structure also have limited use for the interpretation of function within a root system. They are frequently done with material retrieved from soil cores and describe small regions at high resolution. The techniques do not allow an appreciation for the developmental sequences that occur along any root, or for the heterogeneity of structure among roots of an individual plant.

There is, therefore, a large and awkward gap in the root structural information available to those attempting to interpret data relating to root function in field-grown plants.

In recent years my colleagues and I have been attempting to close this gap by applying the techniques and approaches of plant developmental anatomy to the study of the roots of field-grown grasses, particularly maize. This account presents some of our findings with reference to the related work of others.

THE MAIN FRAMEWORK ROOTS

A mature field-grown maize plant without tillers has about 70 nodal roots which, together with the primary root (it usually persists), form the axes of the root system (Hoppe, McCully & Wenzel, 1986). In the variety Seneca Chief, seven tiers of nodal roots develop sequentially beginning at the coleoptilar node. Roots of the 7th tier are aerial and grow into the soil just before maturation of the tassel.

When plants are gently excavated and shaken, two root types are clearly distinguishable. There are long, bare, highly-branched roots whose tips are difficult to locate, and unbranched or weakly-branched roots with white elongating tips and a persistent soil sheath clinging to the rest of the root (Vermeer & McCully, 1982; McCully & Canny, 1985). The sheathed roots are younger than the bare roots. All roots when young have at least a weakly-developed soil sheath and its loss is a feature of maturation. Roots formed later differ, however, from the earlier ones in being larger in diameter (Hoppe, McCully & Wenzel, 1986) and in having more vascular elements and a heavier a more persistent shoil sheath. We have found both sheathed and unsheathed roots on most grasses which we have examined (Fig. 1-3). Similar root dimorphy was first reported by Jackson (1922) in barley. She distinguished early-formed 'branched' roots and later-formed 'unbranched' roots and, although she did not stress the soil sheaths of the latter, she noted the difficulty of washing all the soil particles from them. Sheathed roots have also been described in some desert grasses (e.g. Price, 1911; Arber, 1934) and are features of some other desert monocotyledons (Dodd *et al.*, 1984). In maize, the soil sheath is only weakly (if at all) developed in greenhouse-grown plants (Vermeer-MacLeod, 1982).

The root-soil interface

The root cap of sheathed roots releases cells from its flanks which are surrounded by mucilage which has been secreted by these cells. Soil particles cling to this mucilage (Fig. 13 & 14 of Vermeer & McCully, 1982). The detached cells and mucilage are left behind by the extending root and they lodge among the developing root hairs. Here also soil clings strongly to the root to form a sheath about 1 mm thick which in the smallest maize roots doubles the root diameter. We have measured sheaths of similar or greater thickness in sorghum and in wheat, oat and barley plants where they may occasionally almost quadruple the root diameter (Figs. 1 to 3). In wheat, oats and maize, a cylinder of soil sheath firmly attached to the underlying root epidermis and hypodermis can be slipped off root pieces (Fig. 3 of McCully & Canny, 1985); in sorghum, the cells separate at the endodermis (Fig. 2).

Thin sections cut through the soil sheath along its length show soil, root hairs and the remaining sloughed root-cap cells intermingled with mucilaginous material (Fig. 4). The fine structure and the staining properties of the mucilage are similar to those of root cap mucilage observed on axenic seedlings (Vermeer & McCully, 1982; Miki, Clarke & McCully, 1980).

The detached root cap cells frequently elongate and some persist alive within the soil sheath for many cm behind the root tip. We have noted such detached cells in other grasses (Fig. 5). They are always surrounded by mucilage but we do not yet know if they are continuing to synthesize it, though such detached cells on axenic maize roots are capable of incorporating ^{3}H-fucose (Forsyth, 1980). Soil particles cling closely to the detached cells, frequently obscuring them in whole mounts of scrapings from sheaths.

A constant feature of the sheathed region of the grass roots which we have examined is the presence of numerous root hairs, most of which have knobbly curled regions; occasionally some of these hairs branch at such regions. Soil particles cling tightly to the surface of the curled hairs but not to straight portions even of the same hair (inset, Fig. 4). The clinging of soil particles to distorted portions of root hairs is an old observation (see Haberlandt, 1928) and has generally been attributed to the growing hairs being forced to bypass the particles blocking their path. There is evidence that this is not necessarily so, because unsheathed branch roots regularly bear straight hairs. At the distorted regions of the root hairs soil particles are tightly appressed to the wall, an observation interpreted by early workers (see Haberlandt, 1928) as indicating gelatinization of the wall. Arber (1934) states that, in the sheath-forming desert grass *Aristida pungens*, the root-hair-bearing epidermis secretes the mucilage which cements the sand particles. There is no modern study of localized change in adhesiveness of the root hair surface. Curled and branched root hairs occur in field-grown wheat and their frequency is increased by *Azospirillum* association (Patriquin, Döbereiner & Jain, 1983).

Oades (1978) has reviewed the literature relating to the presence and nature of mucilages on the surfaces of roots. They have been

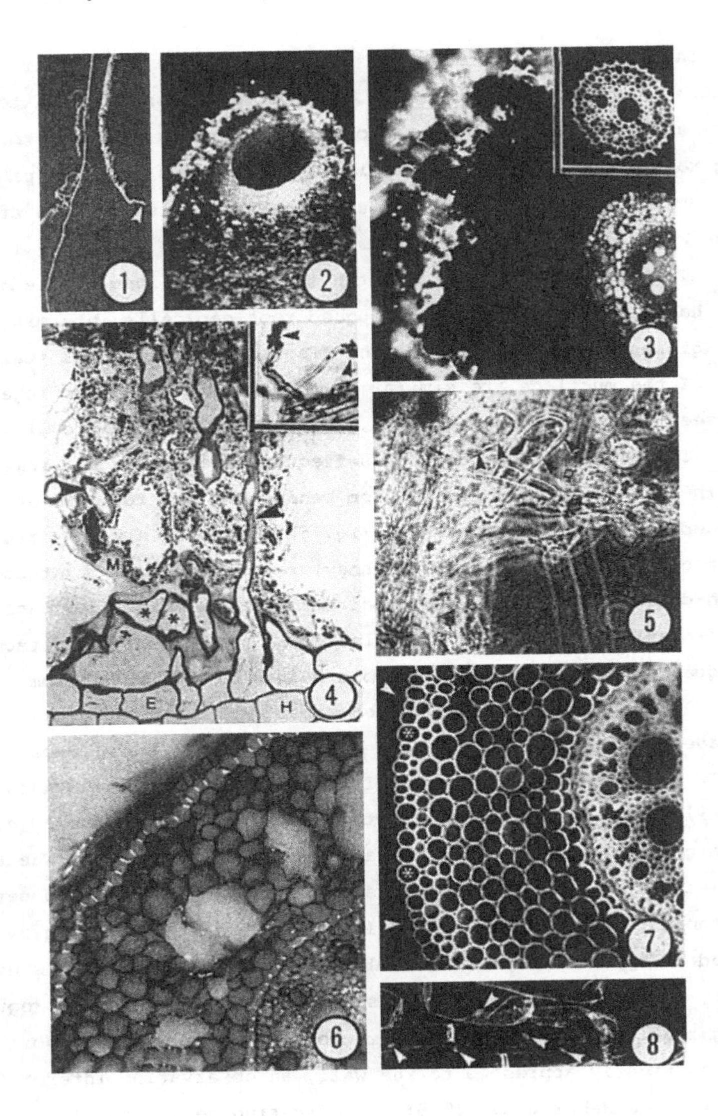

All figures (on pages 56, 57, 62 and 63) are of field-grown roots, except Fig. 16 which shows a root grown in soil in the greenhouse.

Fig. 1: A bare seminal root and a nodal root with persistent soil sheath from a gently-excavated oat plant just beginning to flower. Note the unsheathed growing root tip (arrow). Magnification X 0.5.

Fig. 2: The soil sheath and root tissues centrifugal to the endodermis from a nodal root of a flowering plant of sorghum. Such a cylinder is easily removed by gently pulling at a cut end. Magnification X 12.

Fig. 3: A hand section cut across the sheathed oat root shown in Fig. 1. The section was cut under immersion oil and mounted in oil for photomicrography under semi-dark field illumination. The inset shows a hand-cut section of the seminal root stained with rhodamine B and viewed with a fluorescence microscope. Main figure magnification X 20, inset magnification X 80.

Fig. 4: A 2 µm-thick section of maize sheathed root embedded in plastic. The region shown is similar to that in the hand section of oat root soil interface in Fig. 3. Mucilage (M), sloughed root cap cells (asterisks), root hairs (arrows) and soil particles are resolved. The root epidermis (E) and hypodermis (H) were intact and alive. Section stained with the periodic acid-Schiff's reaction. Inset: Root hairs on a fresh sheathed root which has been thoroughly washed. Soil particles continue to cling to curled portions of the hairs (arrows). Toluidine blue staining. Main figure magnification X 300, inset magnification X 50.

Fig. 5: The root-sand interface in a sheathed root of reed-canary grass (*Phalaris arundinacea*) growing on a dune. Living (streaming), detached root-cap cells (arrows) were present among the root hairs, 5 cm from the nearest root tip. Phase contrast optics. Magnification X 250.

Fig. 6: Hand-cut cross-section 5 mm from tip of sheathed nodal root of maize showing Casparian strips developed in both the hypodermis and the endodermis. Autofluorescence remaining after toluidine blue staining. Magnification X 90.

Fig. 7: Hand section cut across a bare nodal root of maize (from node 2 of a plant just prior to flowering) showing the deteriorated epidermis (arrows), the hypodermis (asterisks), the thickened, lignified walls of the cortex and the stelar parenchyma. A small group of cells in the cortex (upper left of micrograph) are disintegrating to form an air space. Autofluorescence. Magnification X 100.

Fig. 8: Cortical parenchyma in a whole mount of tissue stripped from a sheathed root of corn, 20 cm from the root tip. The tissue was stained for 12 h in an aqueous solution of acridine orange (1/1000, w/v). Nuclei are indicated (arrows). Fluorescence optics. Magnification X 100.

described from a variety of plants, but as Oades notes, a general lack of awareness of these mucilages probably results from their easy loss during tissue preparation. No mention is made of soil or sand sheaths in Oades' review, though both Price (1911) and Arber (1934) suggest that root-derived mucilage is cementing the sand particles of the sheaths they describe on the roots of desert grasses. The failure in the intervening literature to recognize the widespread occurrence of sheaths on soil-grown grasses undoubtedly occurred because, in the relatively few studies where soil-grown material was examined, these sheaths and their binding mucilages were deliberately washed off to facilitate embedment and sectioning of the root tissues.

It is generally considered (see Oades, 1978) that on soil-grown plants the root surface mucilage (termed mucigel in this context by Jenny & Grossenbacher, 1963) is heterogeneous, being of both plant and microbial origin. A portion of the mucilage within the soil sheaths is almost certainly derived from bacteria, some of which when isolated are particularly noteable for their encapsulation (Gochnauer & McCully, unpublished; see also references in Oades, 1978). A heterogeneous plant origin for rhizosphere mucilage (i.e. besides the root cap, also from root hairs, epidermal secretions, lysed epidermal cells, etc.) is frequently cited (e.g. Rovira, Foster & Martin, 1979). Although the secretion by root cap cells is very well documented (see Rougier, 1981), there is no experimental data supporting the other suggested sources of plant-derived mucilages in the rhizosphere.

In contrast to sheathed roots, we have not detected surface mucilages associated with bare roots, except where an isolated clump of soil occasionally continues to adhere to the surface.

Tissue external to the endodermis

It is often assumed that in soil-grown roots the epidermis and the cortex external to the endodermis are shed or destroyed during normal root aging (e.g. Russell, 1977; Foster *et al.*, 1983), and that the endodermis becomes the functional surface (Foster *et.al.*, 1983). Although this may be true for many dicotyledonous roots which have secondary growth (though here the endodermis may also be destroyed), exceptions are known (see Esau, 1977). There is variation in respect to the loss of these tissues in the grasses which we have examined.

In maize, all of the axile roots (usually including the primary root) live until the end of the season. The epidermis remains intact and alive on those roots or portions of roots which bear a soil sheath (Figs. 4 and 6; Vermeer & McCully, 1982; McCully & Canny, 1985). In contrast, the bare roots either have no remaining epidermis, or one which is disintegrating (Fig. 7; McCully & Canny, 1985). In both sheathed and bare roots, the underlying cortex remains intact along the length of the root. Roots, both bare and sheathed, occasionally show disintegration of a few cortical cells, leading to air-space formation (Figs. 6 and 7) but this cell death is localized and similar to that described by Campbell & Drew (1983).

In other grasses which we have observed (e.g. oats, wheat, sorghum and Sudan grass), all sheathed roots have intact epidermal and cortical tissues. But by the time of flowering some of the bare roots, particularly the seminal roots, are decorticated, with the endodermis forming the root surface (inset, Fig. 3). In contrast, bare roots which originate from stem nodes usually retain their cortical tissues. Jackson (1922) shows such a root of barley with an intact cortex.

In maize roots, not only do the cortical tissues external to the endodermis persist, but they continue to differentiate parallel with the primary, secondary and tertiary development of the endodermis. (The differentiation of this latter tissue has been described in detail for the grasses, e.g. Clarkson & Robards, 1975, and will not be dealt with here.)

Just distal to the zone of root-hair development in maize, the epidermis begins to stand out from the underlying cortex because of its strongly autofluorescent cell walls and the colour of their fluorescence in the presence of Rhodamine B or Sudan IV. In the root-hair initiation zone the layer of cells immediately underlying the epidermis develops these same fluorescence properties, so that these two surface layers are sharply distinguished from the rest of the cortex. Similar autofluorescence and staining properties with Sudan IV distinguishing these two layers has been shown in young onion roots (Peterson, Peterson & Robards, 1978).

In older regions of the sheathed roots of maize, the characteristic autofluorescence and fluorescence induced by Rhodamine and Sudan fluorochromes in the epidermis and the underlying cortical layer extend to the walls of cells in one or more adjacent cortical layers (Fig. 2A of McCully & Canny, 1985) and also appear in walls of the cortical cells just outside the endodermis. In older bare roots, walls of all the cortical cells show similar fluorescence properties (Fig. 7; Fig. 3A of McCully & Canny, 1985). At this stage, all cortical cells have also developed thick secondary walls which are lignified. Formation of these thick walls begins in the subepidermal layer and the innermost layer next to the endodermis and extends to the intervening cells. In the oldest roots, one or two layers of cortical cells adjacent to the endodermis have very thick inner tangential walls somewhat like those of the endodermis (Fig. 7). Symplastic connections between cortical cells are retained through deep pits. The walls of the epidermal cells do not appear to undergo any secondary thickening before the cells disintegrate.

Suberized lamellae (sometimes also Casparian strips) in the walls of the cortical cell layer beneath the epidermis in roots of different species have been reported for many years. When these wall modifications are present, the cell layer has been called a hypodermis or exodermis (Esau, 1977). Recently, the development of the Casparian strips and the suberin layers in the hypodermis of onion and maize have been thoroughly documented (Peterson *et al.*, 1978; Peterson, Emanuel & Wilson, 1982; Peterson & Perumalla, 1984). Perumalla & Peterson (1986) have shown that, in maize seedlings, the distance from the root tip at which a Casparian strip develops varies greatly with the growth rate and the age of the primary roots. These Casparian strips always appear at a level proximal to those of the endodermis except when root growth has been strongly inhibited, when both hypodermis and endodermis develop Casparian strips as close as 5 mm from the tip. In the axile roots of field-grown maize, Casparian strips in hypodermal cells appear in the region of root-hair initiation at the same level at which they also develop in the endodermis (Fig. 6). A suberized lamella develops quite soon after the Casparian strip in cells of both tissues.

The Sudan and Rhodamine-induced fluorescence, together with the autofluorescence of the maize root epidermis and hypodermis (and in older roots also of the rest of the cortex), suggests the presence of suberin (see also Peterson *et al.*, 1978), though electron micrographs of the maize roots show a clear suberized lamella only in walls of the hypodermis.

Suberization of the hypodermis near the base of the roots of maize seedlings has been correlated with reduced radial movement of phosphate through the cortex (Ferguson & Clarkson, 1976). Any effect on ion uptake of the suberization of other tissues external to the endodermis, or of their development of thick secondary walls, remains to be investigated. There is, however, evidence for movement of carbon compounds out of these roots into the rhizosphere. Tracing the exudation of translocated ^{14}C from specific regions of field-grown maize roots has shown a similar amount originating from areas of sheathed roots with structure like that shown in Fig. 6 and from unbranched portions of bare roots resembling that shown in Fig. 7 (McCully & Canny, 1985).

The finding of intact cortical tissues on mature field-grown roots of grasses, tissues which in maize are actively involved in carbon and nitrogen metabolism (McCully & Canny, 1985), contrasts with the reported disappearance of nuclei from cortical cells in barley and other cereals (see Henry & Deacon, 1981). Cortical cell death may well be occurring in the roots used by Henry & Deacon, but caution is advised against hasty interpretation of the results of similar staining methods used on old, field-grown roots. The cortical and epidermal cells of maize which have developed the fluorescence properties described above, are almost impermeable to solutions of acridine orange and other dyes. The density of nuclear staining shown in Fig. 8 requires at least 12 h staining of thick sections.

BRANCH ROOTS

The structural aspects of branch-root initiation and early development have been investigated quite extensively, particularly in maize, although nearly always with reference to seminal roots (see the review of McCully, 1975). There are very few studies of the structure of emerged branch roots. The work with field-grown maize has recently focussed attention on the following features, all of which require further investigation.

Size and tissue components

There is a surprising range in the diameters of branch roots in a maize plant even among those roots arising over a short distance from a single axis (Fig. 9). A similar variation, in which no pattern can be seen, occurs in both first- and second-order branches (higher branching orders are rare in field-grown plants) and there is overlap in the diameters of the smallest first-order branches and the largest second-order branches. There is no clue in the literature to what factors may be responsible for such variation in root diameter.

Regardless of their size, most branch roots of maize, even those remaining on the old bare axes, have an intact epidermis bearing root hairs (Fig. 10-12). In the smallest roots, the cortex is reduced to 2 layers, a hypodermis with identifiable Casparian bands, and the endodermis (inset. Fig. 10). In all branch roots the vascular tissue is reduced in extent compared to that of the framework roots of even the smallest diameter, i.e.

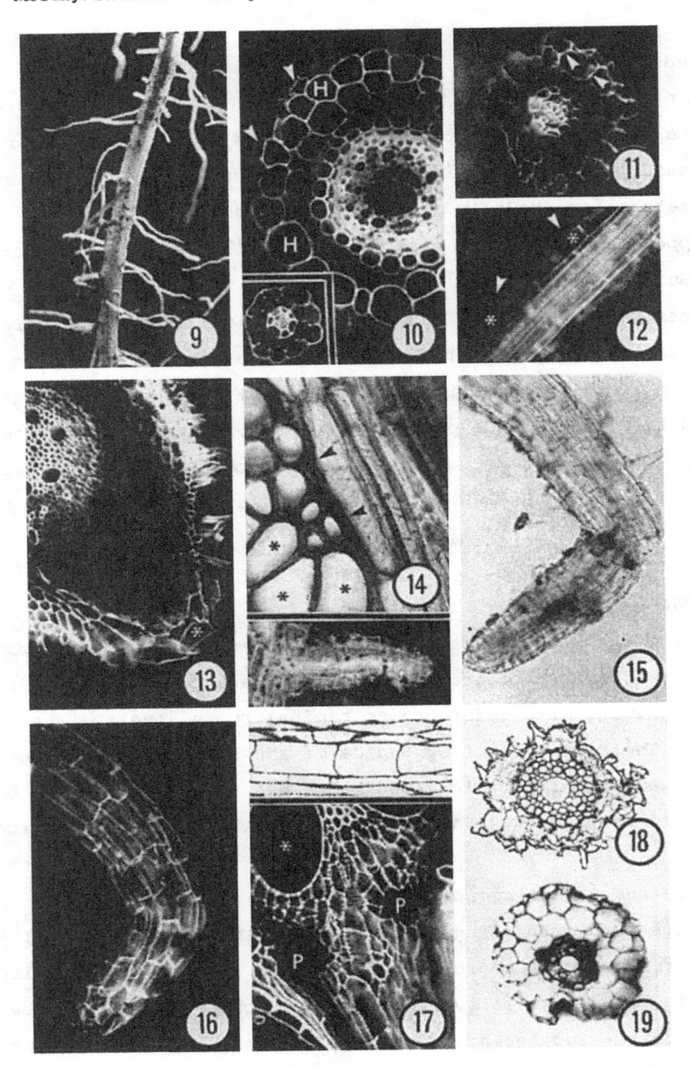

Fig. 9: Typical portion (taken 15 cm from the stem) of a bare root, the primary root of a field-grown maize plant (at the time of flowering). The tipless branch roots are also typical of many of the lateral roots on bare nodal roots. Magnification X 3.

Fig. 10: Hand-cut section of a first-order branch from the primary root shown in Fig. 9. The epidermis (arrows) is still partially intact, though the walls are not thickened. In contrast, most cells of the hypodermis (H) and underlying cortex have lignified secondary walls. The endodermis has a thick tertiary wall and central metaxylem elements are present. Rhodamine B-induced fluorescence. Inset: a second-order branch of this root, viewed the same way. Main figure magnification X 150, inset magnification X 165.

Fig. 11: Hand-cut section of first-order branch of a bare nodal root of maize. The Casparian strips (arrows) in the hypodermis are apparent. Rhodamine B-induced fluorescence. Magnification X 140.

Fig. 12: Whole mount of a branch root similar to that shown in inset of Fig. 10. The intact epidermis (arrows) is just visible. There is a single layer of hypodermis (asterisks) overlying the endodermis. Autofluorescence at pH 8.5. Magnification X 125.

Fig. 13: Hand-cut section across a sheathed nodal root of wheat. The section passes somewhat past the mid-line of a young branch root which has just emerged from the parent root. Enlarged hypodermal (asterisks) and epidermal cells of the parent root form a tight collar around the branch root. Rhodamine B-induced fluorescence. Magnification X 75.

Fig. 14: Hand-cut section across a bare nodal root from the second node of a flowering maize plant. The section passes longitudinally through the epidermis and cortex (right side of the micrograph) of an emerged branch root. Tissues of both parent and branch root have lignified secondary walls and there is a graft-like union (arrows) between the epidermis of the branch and the parent cortex and hypodermis. Cells of the latter tissue (asterisks) have enlarged to form a tight collar around the base of the branch root. Toluidine blue staining. Inset: Whole mount of a second-order determinate branch root on a sheathed nodal root of wheat. Apical meristem and root cap are absent. Main figure magnification X 140, inset magnification X 65.

Fig. 15: Whole mount of a first-order branch root from a bare nodal root of maize. The root tip is determinate with no root cap and with cells differentiating right to the tip. The distal end of this root is breaking off at the point where there was a bend in the root axis. Magnification X 65.

Fig. 16: A second-order root of maize showing determinate growth. Root cap and apical meristem are absent. The epidermal cells right to the tip were alive and streaming. Whole mount viewed with autofluorescence. Magnification X 160.

Fig. 17: Hand-cut section across a bare nodal root of maize showing vascular connections of a branch root. Phloem poles of main root (P) are unstained. Stelar parenchyma has differentiated so that there is a complex xylem link between the branch root and the adjacent large metaxylem elements (asterisk) of the parent root. Rhodamine B-induced autofluorescence. Inset: Plastic-embedded 2 μm-thick section cut longitudinally just behind broken end of a branch root such as shown in Fig. 15. The single file of large metaxylem elements in the centre of the root have retained their cross walls. Main figure magnification X 75, Inset magnification X 165.

Fig. 18: A 2 μm-thick section of a first-order branch on a bare nodal root of maize. This root was fixed in glutaraldehyde, post-fixed in osmium tetroxide and embedded in Spurr's resin by conventional procedure. Toluidine blue staining. Magnification X 165.

Fig. 19: Hand-cut section of a first-order branch root adjacent to the one shown in Fig. 18. This root was fixed whole in glutaraldehyde prior to sectioning. Rhodamine B staining, bright field optics. Magnification X 100.

those from nodes 1 and 2 which average 0.8 and 1.4 mm diameter with 7 and 12 large metaxylem elements respectively (Hoppe, McCully & Wenzel, 1986). In the axile roots, the large metaxylem elements are always in a ring surrounding a central pith (Figs. 2A and 3A of McCully & Canny, 1985); when present in branch roots, these elements (occasionally as many as 3 but usually 1) are central (Fig. 10). The large metaxylem elements are absent from the smallest branches. Roots as small as 70 μm diameter have been seen on field-grown plants (insert, Fig. 10); these were second-order branches. In these small roots the endodermis consisted of 6 cells and the stele included 4 small peripherally-located xylem elements, 2 sieve tubes and about 8 other stelar cells: there was no pericycle. Miller (1981) has examined in detail the dimensions of branch roots and their component cells in hydroponically-grown seminal roots of maize and found a range in root diameters comparable to that of the field-grown material. The variation in xylem arrangement was also similar; in the hydroponically-grown roots there were no large xylem elements in roots less than 100 μm in diameter.

Tiny roots of 70-100 μm diameter are very common in field-grown maize plants (Miller, 1981, reports the rare occurrence of roots as small as approximately 60 μm diameter). Such hair-like roots are quite difficult to see with the unaided eye. Also many will be lost in any excavation procedures which involve vigorous washing. Interestingly, roots in this diameter range have been reported elsewhere only in trees (Lyford, 1975, 1980). For example in red oak, red maple and paper birch, numerous forest floor roots of fourth-order or higher branches are as small as 60 μm diameter.

Connections between branch and parent roots

Branch roots arise endogenously, destroying cortical and epidermal cells in their path. The relative importance of controlled enzymatic hydrolysis and mechanical degradation in this process is still unclear (see McCully, 1975).

Although damage to parent root tissues is remarkably localized (see for example Fig. 22, Bell & McCully, 1970), it has often been assumed that an opening remains around the branch allowing access to the apoplast of the parent root cortex (Esau, 1977). Indeed, such access of tracer dyes has been demonstrated in hydroponically-grown roots (Peterson, Emanuel & Humphreys, 1981). Sections through the base of branch roots in field-grown maize, wheat and oats, however, show that a strong graft-like union forms between the parent cortex and the branch root epidermis. In older roots this union and the participating cells are strongly lignified (Fig. 14). There is also in these grasses a marked response of the parent root epidermis and hypodermis to branch root emergence. Cells, particularly of the hypodermis, enlarge greatly to form a collar which tightly encloses the base of the branch forming a bulge on the parent root surface (Figs. 13 and 14). The effectiveness of this apparent seal in barring access to the apoplast of the parent root cortex remains to be tested. Clearly such a union is likely to be important in restricting invasion by microbes. Endodermal continuity between parent and branch roots is established early in development providing a boundary to the apoplast of the stele which is probably effective by the time the branch root has emerged, though this has never been tested.

The vascular connections between branch and parent roots must play a key role in the functioning of a root system. Yet, as noted earlier (McCully, 1975), very little information is available. Simple hand-cut sections of maize roots reveal that both xylem and phloem connections are extensive, with direct links formed to at least 4 poles of each tissue in the parent root (Fig. 17). One of the important functions of these connections to the branch roots may be the local bridging of the gap between the large and small metaxylem elements of the parent root. This gap, which is a feature of grass roots, is formed by the encirclement of each large metaxylem element by a sheath of living parenchyma at least one cell wide. The direct differentiation of some of these sheath cells into xylem elements at the base of each branch links the large element to the centrifugally-lying small elements at each of the affected poles. The connections between the axile xylem elements may be very important for rapid transfers in the apoplast of the stele.

The phloem connections differentiated at the base of branch roots in maize are particularly extensive. Sieve tubes may be formed directly adjacent to the linking xylem elements and occasionally even from some of the parenchyma cells forming the sheath around a large metaxylem element, thus creating the very unusual direct apposition of xylem elements and sieve tubes.

Determinate growth

One of the most surprising findings from the study of field-grown maize is the almost complete absence of tips on branches and sub-branches of old, bare framework roots (Fig. 9). At first it was thought that the tips were being broken off during collection, but careful dissection of roots in clods of earth brought back to the laboratory showed that this was not so. The detipped branch roots have living cells (as evidenced by the presence of nuclei and streaming cytoplasm) within one or two cell layers of the tip. There are phenolic and suberized deposits in the walls of the dead cells at the end. As well, epidermal cells at the end frequently bear root hairs. The large central metaxylem elements behind the broken tip have intact end walls (inset, Fig. 17), a feature which suggests that these cells are either still alive (though we did not detect cytoplasm in them) or that the retention of the end wall results from an abnormal differentiation sequence. The presence of intact end walls would be advantageous in minimizing air embolism in the root system.

The tips of young branch roots on sheathed framework roots are usually intact and have a normal meristem and root cap development. As these roots age, however, they show a variety of modifications suggesting the onset of determinate form. The root cap is very small or completely absent and cells of the root meristem have differentiated right to the tip, which is often surrounded completely by elongating epidermal cells (inset, Fig. 14; Figs. 15 and 16). Sometimes an enlargement caused by localized increase in cell division and lateral expansion of the cortex occurs behind the root tip, often accompanied by bending. Occasionally tips have been observed which were abscising just proximal to the curved region.

It is still unclear what is causing the outgrowth and disappearance of apical meristems and the abscission of the tips. It is not even certain that the two phenomena are steps in the same development sequence. Some of the structural modifications are certainly consistent with the sorts of damage to roots which can be caused by soil invertebrates (Ruehle, 1973), though the determinate tips are quite unlike nematode-caused galls on grass roots (Griffiths, Robertson & Trudgill, 1982). The determinate tips and the broken ends are, on the other hand, also features of pot-grown maize plants, though here not as widespread as in field material; relatively few nematodes were ever detected in these pots.

There is some recognition in the literature of differing growth potentials among branch roots in different positions on the main root axes of herbaceous dicotyledons (e.g. Charlton, 1966; Yorke & Sagar, 1970) and trees (e.g. Lyford, 1975, 1980; Jeník, 1978; Dell & Wallace, 1981) and even reference to some roots showing determinate growth (e.g. Dell & Wallace, 1981). Tippett (1982) describes the shedding of short branch roots in gymnosperms by formation of a basal abscission layer. Clearly low growth potential is a feature of the many short branch roots found in mature regions of any root system. The structure of their root tips does not appear to have been previously studied in detail, except in the special case of the 'determinate' lateral roots growing out from nodules in some non-leguminous plants (Torrey & Callaham, 1978); these have a greatly reduced root cap and mature vascular tissue close to their tip.

A NOTE ON THE COLLECTION AND PREPARATION OF FIELD-GROWN ROOTS FOR STRUCTURAL STUDY

Excavation methods which involve anything except the most gentle washing of roots are unacceptable because of the information loss from surface features and by breakage of fine roots. Soil cores which can be dissected in the laboratory avoid these problems but lead to the greater drawbacks of disorientation within the root system and possible autolytic changes before fixation.

We have found it useful to plant maize in the field in undisturbed soil enclosed by 1 m^2 bottomless boxes with sides removable in small sections. For collection, trenches are dug along outside the box, pieces removed sequentially from the sides and roots located by horizontal excavation. For preparation of resin-embedded tissues (Vermeer & McCully, 1982), material can be fixed directly in the field or clumps of soil from known orientation returned to the laboratory, dissected under a microscope and the fine roots immediately fixed. But I would like to emphasize the value of hand-cut sections either of fresh (if time permits) or fixed roots, particularly when these are viewed with epifluorescence optics. Not only are such sections easily and quickly made but they are free of the distortions that resin (or wax) embedding may introduce. The cortical tissues of small roots are particularly prone to this distortion, which appears to result from an osmotic effect from resin monomer molecules penetrating the suberized and phenol-rich cell walls much more slowly than do the exchanging solvent molecules. The differences between the apparent structure of similar roots observed fresh or in resin can be dramatic (compare Figs. 18 and 19). The possibility of procedure-induced distortions should be kept in mind when evaluating apparent viability of tissues in sections of roots.

REFERENCES

Arber, A. (1934). *The Gramineae*. Cambridge: Cambridge University Press.

Bell, J.K. & McCully, M.E. (1970). A histological study of lateral root initiation in *Zea mays*. *Protoplasma*, 70, 179-205.

Campbell, R. & Drew, M.C. (1983). Electron microscopy of gas space (aerenchyma) formation in adventitious roots of *Zea mays* L. subjected to oxygen shortage. *Planta*, 157, 350-7.

Charlton, W.A. (1966). The root system of *Linaria vulgaris* Mill. I. Morphology and anatomy. *Canadian Journal of Botany*, 44, 1111-6.

Clarkson, D.T. & Robards, A.W. (1975). The endodermis, its structural development and physiological role. In *The Development and Function of Roots*, ed. J.G. Torrey & D.T. Clarkson, pp.415-36. London: Academic Press.

Dell, B. & Wallace, I.M. (1981). Surface root system of *Eucalyptus marginata* Sm: anatomy of non-mycorrhizal roots. *Australian Journal of Botany*, 29, 565-77.

Dodd, J., Heddle, E.M., Pate, J.S. & Dixon, K.W. (1984). Rooting patterns of sand plain plants and their functional significance. In *Kwongan Plant Life of the Sand Plain*, ed. J.S. Pate & J.S. Beard, pp. 146-77. Nedland, Australia: University of Western Australia Press.

Esau, K. (1977). *Anatomy of Seed Plants*. 2nd edition. New York: John Wiley.

Ferguson, I.B. & Clarkson, D.T. (1976). Ion uptake in relation to the development of a root hypodermis. *New Phytologist*, 77, 11-4.

Forsyth, S.F. (1980). *Autoradiographic examination of the incorporation of tritiated sugars in cells of the root cap and epidermis of corn (Zea mays) and associated mucilaginous secretions*. M.Sc. thesis. Carleton University, Ottawa.

Foster, R.C., Rovira, A.D. & Cock, T.W. (1983). *Ultrastructure of the Root-Soil Interface*. St. Paul, Minnesota: The American Phytopathological Society.

Griffiths, B.S., Robertson, W.M. & Trudgill, D.L. (1982). Nuclear changes induced by the nematodes *Xiphinema diversicaudatum* and *Longidorus elongatus* in root tips of perennial ryegrass, *Lolium perenne*. *Histochemical Journal*, 14, 719-30.

Haberlandt, G. (1928). *Physiological Plant Anatomy*. Translated by M. Drummond. London: Macmillan.

Henry, C.M. & Deacon, J.W. (1981). Natural (non-pathogenic) death of the cortex of wheat and barley seminal roots as evidenced by nuclear staining with acridine orange. *Plant and Soil*, 60, 255-74.

Hoppe, D., McCully, M.E., & Wenzel, C.L. (1986). The nodal roots of *Zea*: their development in relation to structural features of the stem. *Canadian Journal of Botany*. In press.

Jackson, V.G. (1922). Anatomical structure of the roots of barley. *Annals of Botany*, 36, 21-39.

Jeník, J. (1978). Roots and root systems in tropical trees: morphological and ecological aspects. In *Tropical Trees as Living Systems*, ed. P.B. Tomlinson & M.H. Zimmerman, pp. 323-49. Cambridge: Cambridge University Press.

Jenny, H. & Grossenbacher, K. (1963). Root-soil boundary zones seen in the electron microscope. *Proceedings of the Soil Society of America*, 27, 273-7.

Lyford, W.H. (1975). Rhizography of non-woody roots of trees in the forest floor. In *The Development and Function of roots*, ed. J.G. Torrey & D.T. Clarkson, pp. 180-96. London: Academic Press.

Lyford, W.H. (1980). Development of the root system of northern red oak (*Quercus rubra* L.). Harvard Forest Paper /21.
McCully, M.E. (1975). The development of lateral roots. In *The Development and Function of Roots*, ed. J.G. Torrey & D.T. Clarkson, pp. 105-24. London: Academic Press Inc.
McCully, M.E. & Canny, M.J. (1985). Localization of translocated ^{14}C in roots and root exudates of field-grown maize. *Physiologia Plantarum*, 65, 380-92.
Miki, N.K., Clarke, K.J. & McCully, M.E. (1980). A histological and histochemical comparison of the mucilages on the root tips of several grasses. *Canadian Journal of Botany*, 58, 2581-93.
Miller, D.M. (1981) Studies of root function in *Zea mays*. II. Dimensions of the root system. *Canadian Journal of Botany*, 59, 811-8.
Oades, J.M. (1978). Mucillages at the root surface. *Journal of Soil Science*, 29, 1-16.
Patriquin, D.G., Dobereiner, J. & Jain, D.K. (1983). Sites and processes of association between diazotrophs and grasses. *Canadian Journal of Microbiology*, 29, 900-15.
Perumalla, C.J. & Peterson, C.A. (1986). Deposition of Casparian bands and suberin lamellae in the exodermis and endodermis of young corn and onion roots. *Canadian Journal of Botany*. In press.
Peterson, C.A., Emanuel, M.E. & Humphreys, G.B. (1981). Pathway of movement of apoplastic fluorescent dye tracers through the endodermis at the site of secondary root formation in corn (*Zea mays*) and broad bean (*Vicia faba*). *Canadian Journal of Botany*, 59,618-25.
Peterson, C.A., Emanuel, M,E. & Wilson, C. (1982). Identification of a Casparian band in the hypodermis of onion and corn roots. *Canadian Journal of Botany*, 60, 1529-35.
Peterson, C.A. & Perumalla, C.J. (1984). Development of the hypodermal Casparian band in corn and onion roots. *Journal of Experimental Botany*, 35, 51-7.
Peterson, C.A., Peterson, R.L. & Robards, A.W. (1978). A correlated histochemical and ultrastructural study of the epidermis and hypodermis of onion roots. *Protoplasma*, 96, 1-21.
Price, R. (1911). The roots of some North African desert grasses. *New Phytologist*, 10, 328-39.
Rougier, M. (1981). Secretory activity of the root cap. In *Encyclopedia of Plant Physiology*, New Series, ed. by W. Tanner & F.A. Loewus, Vol.13B, pp. 542-74. Berlin: Springer-Verlag.
Rovira, A.D., Foster, R.C. & Martin, J.K. (1979). Note on terminology: origin, nature and nomenclature of the organic materials in the rhizosphere. In *The Root-Soil Interface*, ed. by J.L. Harley & R.S. Russell, pp. 1-4. London: Academic Press.
Ruehle, J.L. (1973). Nematodes and forest trees - types of damage to tree roots. *Annual Review Phytopathology*, 11, 88-118.
Russell, R.S. (1977). *Plant Root Systems*. London: McGraw-Hill.
Tippett, J.T. (1982). Shedding of ephemeral roots in gymnosperms. *Canadian Journal of Botany*, 60, 2295-302.
Torrey, J.G. & Callaham, D. (1978). Determinate development of nodule roots in actinomycete-induced root nodules of *Myrica gale*. *Canadian Journal of Botany*, 56, 1357-64.
Vermeer, J. & McCully, M.E. (1982). The rhizosphere in *Zea*: new insight into its structure and development. *Planta*, 156, 45-61.
Vermeer-MacLeod, J. (1982). *Histochemical, Histological and Structural Study of Field-Grown Corn Roots*. M.Sc. thesis. Carleton University, Ottawa.

Yorke, J.S. & Sagar, G.R. (1970). Distribution of secondary root growth potential in the root system of *Pisum sativum*. *Canadian Journal of Botany*, 48. 699-704.

FUNCTION OF ROOT TISSUES IN NUTRIENT AND WATER TRANSPORT

M.C. Drew

INTRODUCTION

A major function of the roots of land plants is to transfer in a regulated manner, mineral nutrients and water from the soil solution to the remainder of the plant. Research into mechanisms controlling movement of nutrient ions and water in roots continues to be dominated by concern over the properties of plasma membranes, transmembrane carriers or porters, and ATPases. Fluxes of either ions or water across tissues can be quantified in terms of driving forces (gradients of chemical potential) and resistances, but it is an over-simplification to assume that the plasma membrane is always the only or major resistance to these fluxes. There is abundant evidence that a number of other potential resistances to the free movement of materials across the root are laid down by specific tissues during cell maturation, or in response to environmental stimuli. Such structural changes in the tissues of the root may be adaptations that improve plant survival of adverse environmental conditions, but they exert also an overwhelming influence on the effectiveness with which roots function in nutrient and water transport. The first part of this chapter discusses mechanisms by which nutrient ions and water are thought to cross the tissues of the root towards the xylem. The second part considers how the physical environment influences these mechanisms, particularly in terms of root structure, and thereby modifies the effectiveness of the root as an absorbing organ.

PATHWAYS FOR RADIAL MOVEMENT OF IONS AND WATER TO THE ROOT XYLEM

Potential pathways

Both apoplastic (cell wall) and symplastic (cytoplasmic, with movement between cells via plasmodesmata) pathways are potentially available for the transfer of ions and water across the root cortex and stele to the xylem. Additionally, water may flow by a transcellular pathway comprising movement from the cytoplasm and vacuole across the plasma membranes and walls from one cell to the next in series. It is widely accepted that the trans-cellular pathway is of little importance to the movement of nutrient ions: the combined resistance in crossing a large number of cell membranes in series would be much greater than the overall resistance to the flux of ions moving across the intact root from the outer solution to the xylem. There is also direct evidence that plasmodesmata function was a low resistance pathway in solute transfer between cells. Despite much research, the relative contribution of each pathway in the younger tissues of an intact root, before any modifications to its structure, remains obscure.

Apoplast

The apoplastic pathway in primary cell walls is essentially a hydrated polysaccharide gel, comprising water-filled micropores between cellulose microfibrils and cross-linking molecules of pectin and glycoprotein (Läuchli, 1976a). The effective pore diameter is about 10 nm, about 20 times that of a hydrated potassium ion. The apoplast is thought to be freely permeable to solutes and water, hence the term 'free space'. The apoplastic movement of solutes in roots can be observed with the transmission electron microscope (TEM) by following the penetration of electron-opaque salts of La^{3+} or Pb, which do not readily cross the plasma membrane, or by the *in situ* precipitation of Cl^- with electron-opaque Ag^+ (Robards & Robb, 1974; Läuchli, 1976a). For water, estimation of the proportion moving in the apoplast is difficult because the locations of the major resistances to movement across roots are unknown. Measurement of the hydraulic conductivity (L_p) of cell wall preparations yields wide ranges of values. Weatherley (1982) calculates that the conductance of the cell wall of a cortical cell is likely to be 25-250 times greater than that of the plasma membrane. Using an equally plausible argument, Newman (1976) concludes that the apoplastic pathway is likely to give the largest resistance, and the symplasmic the least. An unsatisfactory aspect of such calculations is their sensitivity to the value of L_p assigned to the walls. Since this cannot be estimated *in situ* in radial walls in the direction of flow, values derived from flow *across* cell wall preparations should be regarded with caution.

The Casparian band of the endodermis is a cell wall incrustation of suberin and possibly lignin that blocks the inter-microfibrillar pores. It occurs close to the tip, probably in all healthy roots. In barley it can be identified within 5-7 mm of the tip, and it effectively blocks the continuity of the apoplast, as evidenced by an arrest of inward movement of

La^{3+} at that point (Robards & Robb, 1974). The plasmalemma is tightly bound to the Casparian band so that further movement of ions and water towards the xylem must take place through at least one layer of living cells, the endodermis. Entry into the symplast entails transport across a plasma membrane (not necessarily at the endodermis) and ensures that selectivity between different ions takes place. The inner limit to the apoplast imposed by the Casparian band is also essential both for the development of hydrostatic pressure in the xylem (root pressure) and for conserving the ionic composition of the xylem: without it, water and solutes would simply leak to the outer medium.

During root development, and depending on species and environment, a number of other suberized wall structures are laid down (see later). The general chemical and physical properties of the polymeric material, suberin, are reviewed by Kolattukudy (1980), and its chemical composition and ultrastructure in the hypodermis and endodermis of maize are described by Pozuelo, Espie & Kolattukudy (1984). The major aliphatic components are long chain (C_{16}-C_{26})-hydroxy and dicarboxylic acids; additionally, phenolic compounds (p-hydroxybenzaldehyde and vanillin) are released by chemical treatment of suberin. Although suberin deposited in the Casparian band and as suberized lamallae is believed to function as barriers to apoplastic movement, its supposed properties are not easily demonstrated *in vivo*. The suberin polymer is hydrophobic, and in the TEM is characterized by alternating electron opaque and translucent bands (for example, see Fig. 5), thought to represent layers of polymer and non-polar lipid or wax. Studies with wound periderm of potato tubers (Kolattukudy, 1980; Vogt, Schönherr & Schmidt, 1983) and to a lesser extent the periderm of *Betula* stems (Schönherr & Ziegler, 1980) reveal that the major resistance to water flow is contributed by the wax component. However, the presence of suberin in the cell wall does not necessarily indicate a low permeability in the apoplast, for the deposits may be irregular without forming continuous lamellae, leaving hydrophilic channels for movement of water (Schönherr & Ziegler, 1980).

Movement of ions through the apoplast to plasma membranes will be influenced by the net negative charge arising from ionization of cell wall components (pectins and proteins). At present there is no generalized theory to deal with this complex system. The ionic environment adjacent to plasma membranes is clearly very different from the equilibrium outer solution because adsorbed cations, predominantly Ca^{2+}, Mg^{2+} and H^{+} surround the exchange sites. Calculations suggest that locally the pH may be as low as 3.3 when the outer solution is pH 5.5 (Sentenac & Grignon, 1981). The degree of protonation of carboxyl groups in the wall will influence the amounts of adsorbed cations, selectivity of exchange sites, and ion concentrations close to plasma membranes. Although theory can now successfully describe such exchange properties with isolated cell wall preparations (Sentenac & Grignon, 1981; Demarty, Morvan & Thellier, 1984) the quantitative effects of proton efflux from protoplasts in intact tissues are less easily predicted. From Donnan considerations, the concentration of anions in the cell walls must be greatly reduced, perhaps by 100-fold (Walker & Pitman, 1976), which would make the concentration of some anions like phosphate that only attain 0.1-1.0 mmol m^{-3} in the soil solution, extremely small at the plasma membrane. However, in walls of the aquatic *Enteromorpha intestinalis*, anions appear to penetrate rapidly through pores, thereby avoiding the influence of the fixed negatively charged sites (Ritchie & Larkum, 1982). There is no comparable information yet for root cells. A second consequence of the electrostatic field in the cell walls is to retard the free space diffusion of cations. Pitman (1982) estimates that self diffusion coefficients for cations and anions, approximately 10^{-9} m^2 s^{-1} in free solution, decrease for univalent anions, univalent cations and divalent cations, respectively, to 1-5 x 10^{-10}, 0.5-1.0 x 10^{-10} and 0.3 x 10^{-10} m^2 s^{-1}. Whether this slower diffusion is of any consequence to radial ion transport may depend on the distance across the root cortex that ions move in the apoplast before entry into the symplast.

Symplast

The general properties of the symplast in relation to the movement of ions have been reviewed (Spanswick (1976) and Läuchli (1976b)). More recently, additional evidence for the existence of a low resistance pathway has come from studies of electrical coupling in roots of *Azolla pinnata* (Overall & Gunning, 1982), where the passage of a small current from one cell to another, carried by ion migration, correlates with the number of plasmodesmata in the intervening cell wall. Furthermore, intercellular movement of fluorescent probes (water soluble fluorescent dyes and dye conjugates to which the plasma membrane is impermeable) injected into individual cells allows a direct visualization of symplastic transport (Goodwin 1983). By testing a range of molecular sizes, the upper limit for movement through plasmodesmata in leaves of *Elodea canadensis* corresponded to a relative molecular mass of 874 giving a pore diameter of 3-5 nm. Such a size limit would permit movement of nutrient ions, and metabolites of low molecular mass like sugars, aminoacids and nucleotides but exclude proteins and nucleic acids, thereby maintaining the identity of each cell at the translation level.

Despite improved physiological evidence, knowledge of the structure of plasmodesmata is still incomplete. One difficulty is the interpretation of fixed, static structures at the limits of resolution with current TEM techniques. Calculations (Tyree, 1970) support the view that known rates of symplastic transport through a sequence of cells could be accounted for, quantitatively, by a combination of cyclosis of the cytoplasma and diffusion along plasmodesmata, but the dimensions of plasmodesmatal pores are critical. The earlier model of plasmodesmatal structure proposed by Robards (see Gunning & Robards, 1976), which views the central 'desmotubule' as an open lumen in continuity with the endoplasmic reticulum (ER) of the two adjoining cells, may require revision. TEM studies of plasmodesmata of roots of *Azolla pinnata* (Overall, Wolfe & Gunning, 1982) and lettuce (Hepler, 1982) reveals the desmotubule with an occluded core, and communication between adjoining cells only possible via the cytoplasmic annulus.

The implications for the function of plasmodesmata in symplasmic transport are fundamental. Movement through the symplast to the xylem, based on the Robard's model, would generate a bigger resistance, requiring ions and water to cross not only the plasma membrane but also the ER to enter the symplast. Leaving the symplast at the xylem would necessitate crossing the same two membranes (in reverse order). The alternative model, requiring movement to take place through plasmodesmata in the cytoplasmic annulus, implies that it is the cytosol compartment, not the ER, that constitutes the symplasm, with only half the minimum number of cell membranes (and therefore resistances) to cross. It is relevant to note that in the gap junction of animal cells (a structure equivalent to the plasmodesmata) it is the cytosol compartment that passes across the gap, with a molecular mass exclusion limit similar to that found in plant cells.

An additional feature of the structure of plasmodesmata to consider is the proposed 'sphincter'-like function in the neck region, where the cytoplasmic annulus appears to be narrowed by close contact with the plasma membrane, perhaps by contractile proteins (Olesen, 1979 and 1980). The narrow dimensions observed in the TEM would suggest a prohibitive resistance to flow if water movement were to take place only in the symplasm (Weatherley, 1982). Calculations based on this assumption and on known rates of water flow across the endodermis indicate that the combined resistance of the plasmodesmata in the tangential walls would be ten times the resistance of the plasma membrane. The possibility remains, however, that under a hydraulic gradient induced by rapid transpiration, the apparent neck constriction, or the partial occlusion of the cytoplasmic annulus by electron-opaque particles (Overall *et al.* 1982) might be relieved by dilation, with a large reduction in resistance.

Direct evidence of regulation of movement of solutes through the symplast by plasmodesmata is provided by following the movement of fluorescent probes in leaves of *Egeria densa* (Erwee & Goodwin, 1983). Injection of small quantities of Ca^{2+}, Mg^{2+} or Sr^{2+} into the cell immediately before injection of a fluorescent probe, strongly inhibited movement of the probe to adjacent cells. These Group II ions, and a variety of chemical treatments that would be expected to raise free Ca^{2+} concentrations in the cytosol (CCCP, trifluralin, Ca-ionophore), inhibited symplastic movement in less than 1 min. The effect could not be attributed to cell damage, for with Ca^{2+} additions the inhibition of transport was lost after 5-30 min, whereas comparable additions of Na^{+} or K^{+} failed to elicit any inhibition.

To what extent might transport in the symplast be directed by the location and frequency of plasmodesmatal connections? It seems well established that plasmodesmata originate during cell division, elements of tubular ER becoming enclosed by the developing cell plate (Robards, 1976; Hepler, 1982). However, comparison between plasmodesmatal frequencies in walls of cells in the meristematic zone of roots, with those in slightly older cells in the zone of cell expansion (Seagull, 1983) showed that in the 4 species examined (*Trifolium repens*, *Raphanus sativus*, *Zea mays* and *Sorghum vulgare*) secondary formation of plasmodesmata must have occurred during cell expansion, the new plasmodesmata being clustered around existing ones to form pit fields. The greatest change was in *Zea mays*, where average numbers (μm^{-2}) increased from 3 to 13, with the number per pit field rising from 4 to 9. Secondary formation of plasmodesmata must involve local dissolution of cell walls, and coordination between adjacent cells in the synthesis and assembly of new plasmodesmata.

Because they provide a pathway of low resistance for the cell to cell movement of ions, it is possible that the relative distribution of plasmodesmata in cell walls could exert a marked influence on the pathway for ion movement across the root to the xylem, particularly if the secondary formation of plasmodesmata is sensitive to environmental stimuli. Regrettably, there is little information on the relative frequencies of plasmodesmata in cell walls in mature regions of the root where the major transport of ions and water occurs. In roots of *Raphanus sativus* the number of plasmodesmata (μm^{-2}) in radial walls is only 6-8% of that in the tangential ones in the epidermis or hypodermis before full expansion (Kurkova, 1981), which would strongly direct symplasmic flow in the radial direction towards the xylem. Incidentally, estimates of plasmodesmatal frequency in older cells in Kurkova (1981) may be erroneous because they fail to take account of any possible secondary formation of plasmodesmata.

The function of root hairs and adjoining hairless epidermal cells in the higher aquatic plant *Trianea bogotensis* seems to be determined by the pattern of plasmodesmata connecting each epidermal cell to the underlying cortex (Vakhmistrov, Mel'nikov & Vorob'ev, 1974; Vakhmistrov & Kurkova, 1979; Kurkova, 1981). Net K^+ influx was virtually restricted to the root hair cells, and related to the frequency of plasmodesmata in the inner tangential wall which was some 20-fold greater than in the ineffective hairless epidermal cells. For radish (*Raphanus sativus*) there was no difference in numbers of plasmodesmata between root hair cells and hairless cells, so that there are no grounds for ascribing a specialized absorptive function to root hairs in general. Measurement of the water relations of the root hairs of wheat using a pressure probe show that their hydraulic resistance is similar to that of hairless epidermal cells and cortical cells (Jones *et al.*, 1983), which does not suggest either any specialization of epidermal plasma membranes, or greater plasmodesmatal frequency between root hair cells and the cortex.

The distribution of plasmodesmata in the root pericycle could control the direction taken by materials moving towards the xylem. In the seminal root of barley, 43% of the plasmodesmata were located in the tangential wall between endodermis and pericycle, 45% in the radial walls of the pericycle, but only 12% in the tangential wall between pericycle and stelar parenchyma (Vakhmistrov, 1981). The effect of this pattern of distribution could be to direct materials in the symplast that have crossed the endodermis to move tangentially towards the nearest metaxylem vessel in the outer ring.

The pathway for movement of carbon assimilates from the phloem into the root cortex is unknown. Henry & Deacon (1981) attribute the cortical cell degeneration that is observed in healthy, well-aerated roots of *Gramineae* to the blockage of outward translocation of assimilates by suberization of the endodermis. This proposal neglects the possibility of outward symplasmic movement in plasmodesmata. Evidence from studies of translocation of asymmetrically labelled [^{14}C](fructosyl)sucrose from the phloem into cells of growing roots of maize (Giaquinta *et al.*, 1983), indicated that a symplasmic pathway must have been taken; otherwise, cell wall invertases would have hydrolysed any free space substrate and altered the labelling pattern. It is not easy to envisage how, based on current models of their structure (e.g. Overall *et al.*, 1982), an outward flow of sugars could take place against the inward flow of water in a transpiring plant. To postulate a specialization of plasmodesmatal function, some for inward and some for outward flow, implies a greater degree of structural organization in plants than experiments have yet demonstrated directly.

Entry to, and exit from, the symplast

Various hypotheses accounting for the radial movement of ions in the symplast, and eventual loading of the xylem have been examined in detail (Hanson, 1978; Bowling, 1981; Pitman, 1982). The 'two pump' hypothesis, which requires active fluxes of some ions at the plasma membrane of epidermal or cortical cells (symplast loading) and also at the xylem parenchyma (symplast unloading or xylem loading) is now widely supported. Hanson's (1978) adaptation of the chemiosmotic theory provides a conceptual framework in terms of driving forces and resistances, inward ion fluxes being driven by H^+-coupled entry to the symplast, and a H^+-coupled exit at the inner plasma membrane of xylem parenchyma cells. Experimental evidence points increasingly to the location of an electrogenic ion pump in the stele, presumably at the xylem parenchyma (Okomoto, Ichino & Katou, 1978; de Boer, Prins & Zanstra, 1983). Perfusion of the xylem of excised roots of onion with solutions of known pH (Clarkson, Williams & Hanson, 1984) reveals a close regulation of H^+ flux into or out of the xylem coupled in some way to K^+ and SO_4^{2-} efflux to the xylem. Such physiological evidence accords with the ultrastructure of xylem parenchyma cells, which is characteristic of metabolically active, secretory cells (Läuchli *et al.*, 1974a), and with the more negative membrane potential differences measured in xylem parenchyma cells than in the other cells of the root, which suggests a specialized transport function (Dunlop, 1982).

Whether ions enter the symplast at the epidermis or deep within the cortex is difficult to determine unambiguously. The root cortex has long been regarded as a collecting system for ion entry into the symplast, partly based on the assumption that radial transport to the xylem is limited by the flux across plasma membranes. An alternative view, that uptake into the symplasm occurs almost exclusively at the epidermis, has gained increasing support recently (Grunwaldt *et al.*, 1979; Iren & Boers-Van der Sluijs, 1980; Kochian & Lucas, 1983). Using impermeant sulfhydryl reagents as inhibitors of ion transport at the plasma membranes of maize epidermal cells, Kochian & Lucas (1983) concluded that K^+ uptake from dilute solution (0.2 mol m^{-3}) took place at the outer surface of the root. The notion of a Casparian band in the hypodermis of maize and other species would seem to support this view, although evidence on its function is not conclusive (see next section). In aerenchymatous roots, the formation of gas-filled lacunae takes place by lysis and disappearance of 80% of the mid-cortex. Despite this loss of cortical cells, transport of K^+ or inorganic orthophosphate (P_i) over wide ranges of concentration are as rapid as in ordinary non-aerenchymatous roots (Drew *et al.*, 1980; Drew & Saker, 1986). On first inspection this suggests that the epidermis or outer cortex are the location of entry into the symplast in ordinary roots, but equally, ions seem free to migrate in aerenchymatous roots along the apoplastic pathway constituted by the radial strands of the cell walls that remain after cell lysis. If the endodermis remains directly accessible to ions from the outer solution, initial uptake into the symplast might then take place at this cell layer. The important feature seems to be that, whatever the pathway to the xylem, trans-cortex transport is not rate-limiting.

Estimation of the proportion of water moving through the symplast, or through the transcellular pathway, presents special difficulties, in part because of the lack of a convenient method of tracing water flow on a fine scale. The proposed transcellular pathway would entail movement across a series of membranes, namely tonoplast and plasma membrane at each cell, so that the calculated overall resistance in passing across the root would be unrealistically large (Weatherley, 1982). Recent measurements with barley (Steudle & Jeschke, 1983) and wheat (Jones *et al.*, 1983) challenge that conclusion. The hydraulic resistance of the entire root was about 10 times greater than the hydraulic resistance of individual epidermal and cortical cells measured *in situ* with a pressure probe. Thus, if water enters cell protoplasts at the epidermis, the resistance to water movement across a sequence of 5 to 6 cells (some 10 to 12 membranes) is similar to the hydraulic resistance of the entire root. However, even that conclusion is uncertain because estimates of L_p made with the pressure probe assume that plasmodesmata play no part in conducting water out of the cell under test, an assumption that remains to be established experimentally. At the other extreme, the consequence of a low hydraulic resistance in the apoplastic pathway would be to locate the first major resistance to water movement at the plasma membrane of the endodermis, where the Casparian band blocks the apoplast. If this were so, the endodermis would take on a special role as a water-absorbing tissue, and the L_p of the endodermal plasma membrane would dominate. Indeed, some have assumed that the resistance to water flow across intact root systems is simply equivalent to the resistance of the endodermal plasma membrane. A scheme of turgor-controlled changes in endodermal hydraulic resistance has been put forward based on this assumption (Powell, 1978).

Zones of ion and water transport along roots and changes associated with structural development

Measurements of ion accumulation and transport to the xylem, in which labelled nutrients are supplied to discrete segments or zones along the root axis, have revealed characteristic patterns of activity that appear to be related to structural changes, particularly in the endodermis and hypodermis. Most nutrient ions are strongly accumulated by cells in the apical 5 mm of the root, with little immediate transport to the xylem. However, in more mature tissue, transport of K^+ and P_i to the xylem can be almost equally effective in zones located 10-500 mm from the tip in seminal roots of barley and maize and in the primary root axis of *Cucurbita pepo* before extensive cambial activity (Harrison-Murray & Clarkson, 1973; Robards *et al.*, 1973; Clarkson, 1974; Ferguson & Clarkson, 1975). It is assumed that these ions move in the symplasm, crossing the endodermis by plasmodesmata and thereby avoiding the apparently formidable barrier presented initially by the suberized lamellae that totally enclose the protoplast in endodermal cells (State II endodermis) and then by the additional cellulosic and sometimes lignified thickenings laid down inside the suberin lamellae, mainly on the inner tangential wall (State III endodermis).

By contrast, transport of Ca^{2+} to the xylem reaches a maximum in the zone 50-100 mm from the root tip in barley and *Cucurbita*, and then declines with the transition of endodermal cells from State I to State II and III (Harrison-Murray & Clarkson, 1973; Robards *et al.*, 1973; Clarkson, 1984a). Little symplasmic transport might be expected for Ca^{2+} because of the low concentration of free ions in the cytosol (0.1-1.0 mmol m^{-3}). Clarkson (1984a) argues that most net flux of Ca^{2+} towards the xylem in younger root zones begins at the plasma membrane of State I endodermal cells. Once suberin lamellae isolate the endodermal plasma membrane from direct access to the apoplast (State II), movement of Ca^{2+} across the endodermal plasma membrane becomes limited to a small, residual, symplasmic component.

Changes in water movement to the xylem closely parallel those of Ca^{2+}. By attaching micropotometers to different zones along the root, water movement was observed to decrease to about 10% of the maximum rate during the formation of State II and III endodermal cells in barley (Clarkson, 1984a). This residual water flux has long been thought to represent a symplasmic component, crossing the endodermis via plasmodesmata, on the assumption that the formation of suberin lamellae block access from the apoplast to the plasma membrane of endodermal cells.

A different pattern of water flux in relation to root structure is evident in primary roots of intact maize plants grown at either 13 or 20 ^{o}C (Stephens, 1981). Micropotometers failed to show any decline in water flux in zones up to 15 cm from the tip, where all endodermal cells would have been in State II or III. On the assumption that the suberized lamellae of State II cells would block the apoplast, the flux must have been predominently in the symplast.

In adventitious (nodal) roots of barley, the transition from State I to State II is less synchronized than in seminal roots, leaving 'passage cells' temporarily unsuberized to provide what is thought to be a less resistant pathway for divalent cations and water (Robards *et al.*, 1973).

Selective ion transport and root resistance to water flow depend directly or indirectly on the properties of the endodermis, but there are some imperfections in the system. Where lateral roots initiated in the pericycle extend through the cortex, the endodermis is temporarily ruptured, allowing an unrestricted entry of materials before continuity of the endodermis of main axis and branch roots is re-established (Peterson, Emanual & Humphreys, 1981). An additional 'imperfection' may exist in older basal zones of roots (Sanderson, 1983b) where transport of Ca^{2+} (apoplastically?) is speeded up under conditions of rapid transpiration in a way that is not matched simultaneously by P_i, which probably moves symplastically. Sanderson

(1983a,b) suggests that an apoplastic pathway for Ca^{2+} (and water) movement could exist across State II and State III cells, either where the tight junction between the plasma membrane and Casparian band is displaced by suberin lamellae, or possibly along the lamellae themselves. An alternative interpretation (Sanderson, 1983a) of the decline in water flux in mature zones of the barley root is that it is a consequence, not of suberized lamellae being laid down, but of the thickened tertiary wall. During rapid transpiration there is an approximate doubling of water flux only across the mature zones where the endodermis is in State III, which is attributed by Sanderson to the opening of an apoplastic pathway across the endodermis. However, the dilemma in precisely correlating structure with function in the barley endodermis is not only that there is a lack of synchronization in development of the endodermis, but also that the transition from State I to II in each cell is immediately followed by transition from State II to III. If there is appreciable movement of water (and ions) across the root by a totally apoplastic pathway under conditions of rapid transpiration, as Sanderson maintains, the loss of selectivity in ion transport to the xylem that would occur is difficult to explain in relation to the supposed function of the endodermis.

In many species, the layer of cells immediately beneath the epidermis, the hypodermis (sometimes referred to as exodermis after sloughing or decay of the epidermis in older roots) differentiates to produce suberin incrustations of the cell walls reminiscent of the suberization of State II endodermal cells. The influence on movement of ions and water seems to vary greatly with species and with the environmental conditions prevailing during deposition of the suberin, so that it is not possible to deduce on the sole basis of the occurrence of suberin, the properties in relation to transport. In adventitious roots of onion (*Allium cepa*), suberin lamellae are deposited in the hypodermis and are already prominent 100 mm from the root tip (Peterson, Peterson & Robards, 1978), in a zone which coincides with the maximum for water absorption. Studies with enzymically isolated hypodermal 'sleeves', (Fig. 1) perfused with solution, demonstrate however a high permeability to inward (radial) diffusion of labelled calcium or phosphate as well as to water (Clarkson *et al.*, 1978). Since plasmodesmatal frequencies in the hypodermal walls are very low, the low resistance to ions and water movement presumably indicates the presence of hydrophilic pores in the suberized wall.

In maize, seminal and nodal roots develop hypodermal suberization in older basal zones but the influence on transport varied greatly in different experiments. In some, the suberized hypodermis effectively blocked further inward movement of labelled phosphate (Ferguson & Clarkson, 1976) while in others there was no restriction to water flow (Stephens & Clarkson, 1981). However, using hypodermal sleeves prepared from adventitious roots of maize either exposed to humid air above a nutrient solution, or immersed in it, Clarkson (1984b) showed that the radial permeability was very low in only those zones of the root previously exposed to air. Thus, the chemical or physical properties of the suberin lamellae deposited in the walls of the hypodermis vary in some unknown manner with the physical environment. Roots of rice (*Oryza sativa*) characteristically begin to develop a suberized hypodermis, and a prominent sclerenchymatous layer of lignified cells interior to the hypodermis within a few mm of the root apex (Clark & Harris, 1981). Despite these structures, measurements of labelled K^+ and SO_4^{2-} uptake by discrete root segments on intact plants show no evidence of uptake being

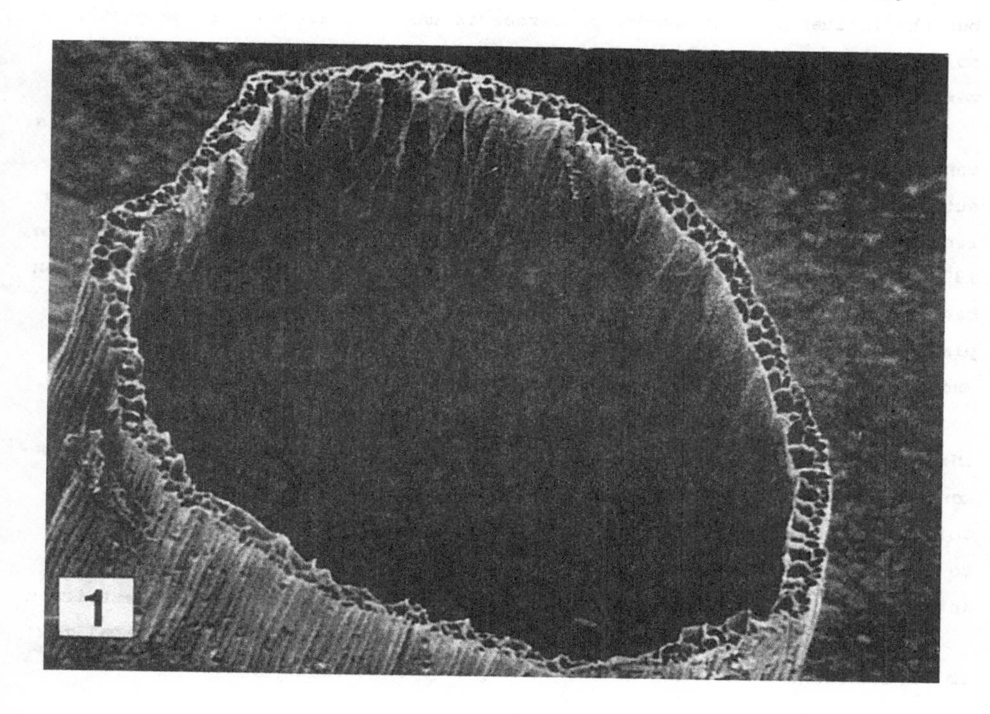

Fig. 1: Epidermal/hypodermal 'sleeve' prepared by enzymic digestion of onion root. SEM, transverse section, bar = 0.5 mm. (micrograph from A.W. Robards, University of York.)

confined to the apical region (Vanden Berg & Clarkson, personal communication). However, plasmodesmata were observed in the infrequent pits between the hypodermis and the sclerenchyma layer (Clark & Harris, 1981), so that symplasmic transport may continue until the cells of the sclerenchyma lose their cytoplasm at maturity. In roots of the sand sedge (*Carex arenaria*), suberized lamellae are found in the hypodermis only 20-50 mm behind the tip. There are few plasmodesmata linking it to other cell layers, and it is highly impermeable to movement of labelled water, P_i or K^+ (Robards, Clarkson & Sanderson, 1979). It is possible in this species that much of the ion and water transport to the xylem is restricted to the fine lateral roots, although this remains to be confirmed.

Suberized lamellae in the root hypodermis have also been identified in *Eucalyptus* species (Tippett & O'Brien, 1976; Dell & Wallace, 1981), and in seagrass (*Posidonia* species; Kuo, McComb & Cambridge, 1982) but the influence on transport processes is unknown, although in seagrass, no plasmodesmatal connections were found between the hypodermis and the next cortical layer.

A structure which resembles the Casparian band of the endodermis can be distinguished in the radial walls of the hypodermis on the basis of autofluorescence and staining properties. It is first detected some 40-50 mm from the root tip in maize, and 30-40 mm in onion (Peterson, Emanual & Wilson, 1982; Peterson & Perumalla, 1984). The significance of a hypodermal Casparian band to radial movement of ions and water depends greatly on whether the plasma membrane is tightly connected to the Casparian band, as it is in the endodermis. Evidence that the band blocks the apoplast was obtained with apoplastic fluorescent dyes, and the possibility of symplastic continuity through the hypodermis at that stage in its development was shown by the movement of uranin (Peterson & Perumalla, 1984). Whether the pathways available to the relatively large dye molecules are representative of those available to ions and water requires confirmation. If it is, then the site of uptake into the symplast in these and perhaps many other species, must be restricted to the epidermis and outer walls of the hypodermis, except in the apical few cm of the root.

THE PHYSICAL ENVIRONMENT, ROOT STRUCTURE AND ROOT FUNCTION

This section considers how the primary physical factors (temperature, water potential, gaseous atmosphere, mechanical resistance) and some chemical and biotic factors, influence the movement of ions and water across the root to the xylem. Here we are concerned with changes in root properties brought about or manifest at the tissue level of organization. It will be necessary to speculate often as to the probable consequence of changes in root structure on ion and water movements because relatively few studies have attempted to relate environmentally-induced physiological and structural changes in roots to their transport function.

Under natural conditions, roots frequently experience diurnal and seasonal variations in temperature and water potential. Mechanical resistance to root penetration and aeration status of different soil horizons can also vary with season and soil hydraulic properties. The manner in which the roots of sensitive species acclimate to adverse soil environments is little understood, but laboratory experiments under controlled environments reveal something of the mechanisms by which the primary physical factors, individually, exert their effect.

Temperature

When chilling-sensitive plants that are acclimated to growth at temperatures close to their optimum are suddenly subject to rapid cooling below a 'critical' temperature, ion and water transport are strongly inhibited (reviewed in Kramer, 1983). This response can be caused by changes in the properties of cell membranes as well as by changes in root structure. At the cell membrane level, there is abundant evidence that the maintenance of the fluidity of the lipid bilayer (avoiding phase transitions to a more ordered structure - freezing - that would disrupt normal membrane properties) is an important feature in temperature acclimation (Lyons, Raison & Steponkus, 1979). Changes in sterol content and the degree of unsaturation of fatty acid side chains in phospholipids are strongly implicated (Clarkson, Hall & Roberts, 1980; Osmond, Wilson & Raper, 1982; Uemura & Yoshida, 1984). Evidence of a reversible acclimation of ion and water transport mechanisms comes from studies with rye (*Secale cereale*) and barley precooled for 2 d at a temperature of 8°C (Clarkson, 1976). Ion uptake and xylem exudation were stimulated 2 to 3-fold by precooling compared with roots maintained at 20°C, but the effect was reversed after returning plants to 20°C for 24 h.

Further analysis of the factors controlling the greater fluxes through precooling has proved elusive. Although low temperature in rye led to an increased desaturation of phospholipids (Clarkson, *et al.*, 1980), the pattern of this change was not closely associated with the induction of more rapid transport because the effect on transport, but not phospholipid desaturation, was reversed at 20°C. Changes in fatty acid composition in roots of soyabean also failed to relate to the net rate of NO_3^- uptake, the latter being unaffected by temperature in plants grown between 14 and 22°C (Osmond *et al.*, 1982). Changes in the number or properties of transport proteins in cell membranes might be anticipated at temperatures below some critical level, but even in chilling-sensitive maize (cv LG11) and marrow, no differences could be detected in freeze-fractured faces of cortical cells in terms of the number or distribution of intramembranous particles (IMP), many of which are likely to represent the transmembrane proteins that act as transporters (Robards & Clarkson, 1984).

Sudden changes in temperature slow root growth so that the endodermis becomes suberized (State II and State III) closer to the apex. In roots of bean this structural change is associated with a reduction in xylem exudation rate (Brouwer & Hoogland, 1964). However, in maize plants grown continuously with the roots at lower temperatures (10, 13, 16°C) than the shoots (20°C), the suberization of the endodermis and hypodermis did not diminish water flux (Stephens, 1981). On the contrary, the integrated hydraulic resistance of the roots at the lower temperature was only one-third of that in roots grown at 20°C, thereby compensating for the small size of the root system, and maintaining the rate of transpiration. Presumably flux of water across the endodermis was symplasmic, if we accept the postulate that the suberized lamellae block apoplastic movement.

Water potential

In the field, different parts of the same root system are invariably subjected to very different water regimes. Superficial roots may be desiccated for extended periods while roots in the subsoil can be in contact with 'available' water. Under arid conditions the entire root system can be exposed to severe desiccation. The ability of the growing root tip to withstand water stress is likely to depend on different properties from those required by older non-growing parts of the root system, where an ability to resume absorptive function when the soil is rewetted may be of greater significance to plant survival.

It seems unlikely that roots continue to absorb at water potentials below the conventional wilting point (-1.5 MPa), not necessarily for physiological reasons but because of a loss of soil-root contact and the greatly restricted mobility of ions and water in dry soil. Cellular dysfunction does occur, however, in both desiccation-tolerant and -intolerant species once desiccation leads to disorganization of cell ultrastructure (Bewley & Krochko, 1982). An important property in desiccation-tolerant cells is the ability to repair and resynthesize structures when rehydrated, provided desiccation has not gone beyond a critical level. When root tips of wheat were exposed to dry air until the tissue was in equilibrium with a water porential of -2.4 MPa (Cole & Alston, 1974), the ability to take up ^{36}Cl on rewetting was initially greatly inhibited. Uptake was fully restored after 18 h rehydration, but the authors did not examine the changes in cell ultrastructure that must have accompanied recovery. However, in a TEM study of the root apex of maize, in a zone 600-800 μm from the tip, cells tolerated up to 60-70% loss of water content without irreversible damage following rapid dehydration over solutions of known water potential (Nir, Klein & Poljakoff-Mayber, 1969). With less than lethal dehydration, the disorganized plasma membranes, mitochondria, plastids and nuclei were all capable of re-assembly when re-hydrated, but no such repair followed lethal dehydration (>70% water loss).

The stage in cellular disorganization at which intercellular transport is disrupted has not been characterized. Shrinkage and folding of cell walls undoubtedly exert mechanical stress and Nir *et al.* (1969) also observed separation of the plasma membrane from the cell wall. In the coleoptile of oat (*Avena sativa*) protoplast shrinkage induced by plasmolysis ruptured plasmodesmata (Drake, Carr & Anderson, 1978). Following the recovery of cells to full turgor, symplasmic transport measured by electrical coupling using microelectrodes was severely disrupted.

When roots are exposed to non-lethal water stress over longer periods, particularly when the stress is applied gradually, root growth slows or stops and endodermal or hypodermal tissues may become suberized close to or at the root tip. In response to seasonal drought, roots of pine (*Pinus halepensis, P. resinosa*) cease extension and become 'metacutinized': that is, a suberized layer, continuous with the endodermis, is laid down in the root apex, including the root cap (Wilcox, 1968; Leshem, 1970). The root remains dormant for long periods and it is reasonable to assume, from the known properties of suberized lamellae, that lethal loss of water from the root tip is avoided. Following rehydration the roots can resume extension within a few days.

For some arid-zone species that resist long periods of desiccation, the roots located in the surface soil layers are able not only to resist dehydration but also to resume absorptive functions within a short period of rewetting by shallow penetrating showers (Drew, 1979). An interesting question concerns the way in which suberin retards water loss, yet allows the re-establishment of continuous films of water across the root when water becomes available. Nobel & Sanderson (1984) examined water uptake and water loss by roots of *Agave deserti*. When detached roots were dried at 20°C and 50% relative humidity for 3-6 h, the resistance to inward radial water flow increased 10 to 100-fold. With re-wetting there was considerable recovery of conductivity which in some roots was complete in 6 h. The outward rate of water movement of detached roots decreased about 200-fold during air drying for 3 d, suggesting a gradual blockage of apoplastic pathways for water loss. After 6 h rehydration, the same initially desiccated roots took up water at an appreciable rate. The authors drew attention to the 'rectifier-like' property of roots in retarding outward loss of water to dry soil, while allowing inward movement from wet soil. The mechanism of these reversible changes was not examined, but roots of *Agave* are known to have a periderm. It is noteworthy that analogous properties are found in the periderm of Betula stem (Schönherr & Ziegler, 1980).

A structural feature that may behave in a rectifier-like manner is the hypodermis of *Hoya carnosa*. In aerial or soil-grown roots, the hypodermis differentiates into two cell types (Olesen, 1978). The 'long cells' form suberin lamellae early in their differentiation and after formation of the tertiary wall, lose their cytoplasm. The 'short cells' are characterized by the development of a thickened wall or 'cap' (3-7 μm diameter) in which labyrinthine sets of tortuous, interconnected channels are interspersed with amorphous lignified material (Figs. 2, 3 and 4). One or more prominent pits per cell cross the cap, and plasmodesmata connect to the epidermal cells, presumably maintaining a symplasmic pathway. Olesen (1978) discusses the possibility that with desiccation the labyrinthine channels lose water and become compressed by the lignified masses, eventually closing off the channels and preventing further water loss by the apoplastic pathway. Rehydration would lead to swelling of the walls and re-opening of channels.

Non-woody roots of many species adapted to arid environments form 'sheath roots' (Drew, 1979; Wullstein & Pratt, 1981) in which the lysis and disappearance of cortical cells resembles the aerenchymatous structure induced by oxygen shortage. It has been assumed that the 'sheath' (the persistent epidermis and hypodermis) helps retard water loss but there have been no direct studies of the function of such roots, or whether they resume absorption on rehydration. However, work on epidermal/hypodermal sleeve of *Carex arenaria*, already referred to, suggests that at least some sheath roots have lost their absorptive function. There is a comparable lack of information on the roots of monocotyledonous mesophytic species, including many crops, that experience loss of the root cortex through normal cell senescence in older zones of the root (see also Chapter 3 of this book). In

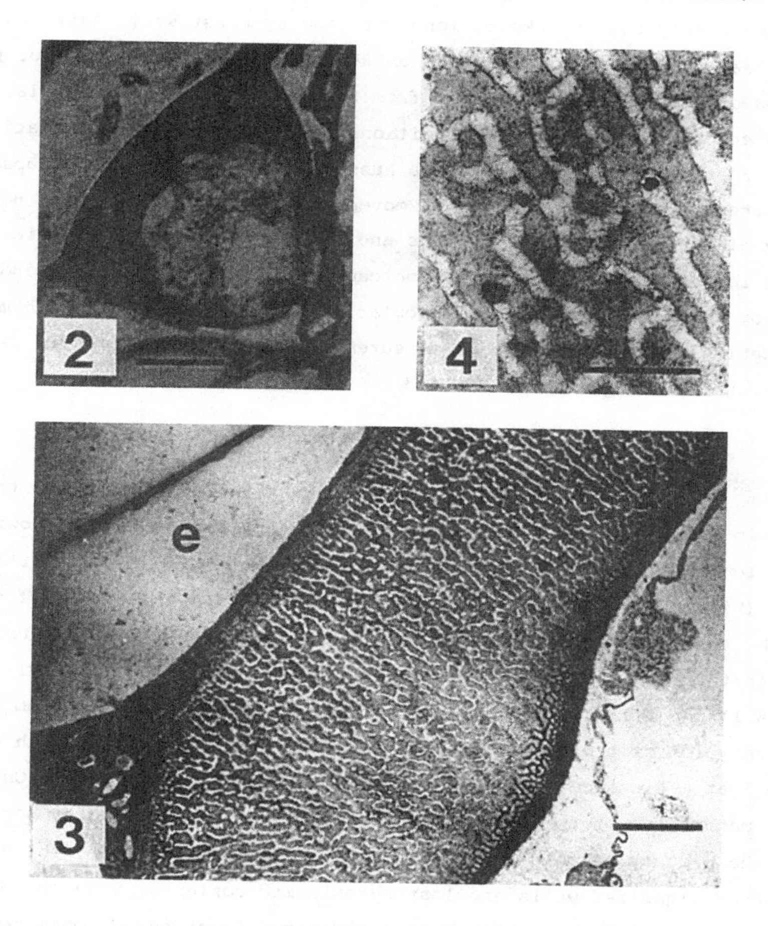

Figs. 2-4: Ultrastructure of hypodermis in *Hoya carnosa*. Fig. 2: 'short cell' with cap (c) and adjacent dense cytoplasm. LM, bar = 10 µm. Fig. 3: cap of short cell showing labyrinthine structure. e, epidermal cell. TEM, bar = 2 µm. Fig. 4: tortuous channels (80-100 nm across) separated by amorphous, lignified material. TEM, bar = 0.5 µm.
(from Olesen (1978), with permission of Springer-Verlag.)

a study of the resistance to desiccation of the older zones of intact barley roots, exposure to dry air for 10 d followed by 3 d rehydration there was a resumption of $^{32}P_i$ uptake and transport (Shone & Flood, 1983). During desiccation the epidermis and outer two-thirds of the cortex collapsed and there was extensive secondary thickening of the endodermis and stele. Barley roots do not form a suberized hypodermis. In these desiccated barley roots it seems likely that uptake of ions into the symplast would have occurred at or near the endodermis rather than at the collapsed epidermis or cortex. Perhaps outward movement of water from the stele maintained the viability of the endodermis under these conditions. With more severe desiccation of barley roots (Clarkson, Sanderson & Russell, 1968) the total collapse of the cortex arrested further radial movement of ions to the stele. However, the axial (upward) movement of ions and water from lower (more apical) zones of the root in contact with solution can continue. The suberized and lignified walls of cells external to the vascular tissues seem to protect them from desiccation, although no direct measurements of net water loss in relation to structural changes have been made.

Mechanical impedance

Resistance of soil to displacement and compression by the growing root tip results in a stress on cells in the apical zone and a slowing of extension. Small, controlled stresses on the roots of barley in artificial media (Wilson & Robards, 1978) showed that despite a shortening of cell length, there was an increase in root diameter due to radial expansion of some cortical cells. In the endodermis the transition from State I to States II and III occurred much closer to the root apex. At 15 mm from the tip, only 13% of the cells were still in State I, contrasting with 40% in State I at a distance of 100 mm from the tip in unimpeded roots. Unusually early maturation of the pericycle took place in stressed roots, beginning with the protoxylem pole cells only 2.5 mm from the root tip, which developed thickened lignified walls and lost cytoplasmic contents. Such changes are likely to restrict transport of ions and water in stressed roots to the apical few mm, since no symplasmic continuity across the endodermis and pericycle seems to remain in the older zones.

Aeration

Although poor gas exchange between the soil and atmosphere is associated with numerous changes in the physical and chemical properties of the soil that influence root function, the most immediate effect on plants is the lack of molecular oxygen for aerobic respiration (see Jackson & Drew, 1984, for a review). When the roots of plants that have not been acclimated to low partial pressures of oxygen are exposed to oxygen-deficient surroundings, inhibition of aerobic respiration and a declining energy status of cells quickly lead to cell injury (Saglio, Raymond & Pradet, 1980; Roberts *et al.*, 1984). A greatly increased resistance to the radial flow of water across the root is experienced within minutes or hours (Mees & Weatherley, 1957; Parsons & Kramer, 1974; Everard & Drew, 1983) much like that induced by respiratory inhibitors. After approximately 10 h anoxia, radial resistance to water flow often declines, probably as a result of gross degeneration of plasma membranes, and there is little resistance to water flow until tyloses formed by cells adjacent to the xylem expand into the lumen and block it (Everard & Drew, 1983). Changes in ion transport are associated with changes in the energy status of the cells. Active transport of K^+ (Cheeseman & Hanson, 1979) or nitrate (Rao & Rains, 1976) are curtailed almost at once under anoxia. Following cell death in the root, which may require 1-7 d of anoxia, depending on species and temperature, movement of ions through the root to the shoot seems to take place largely by mass flow. The average (calculated) concentration of ions in the xylem sap in wheat then becomes similar to that in the outer solution (Trought & Drew, 1980).

For salt-sensitive glycophytes, exclusion of Na^+ or Cl^- from the leaves depends on resorption of ions from the xylem sap by xylem parenchyma cells in the root (Läuchli, 1984). In maize, the mechanism of resorption is inhibited by anoxia so that abnormally large amounts of Na^+ and potentially injurious ratios of $Na^+:K^+$ move to the leaves (Drew & Dikumwin, 1985; Drew & Läuchli, 1985). Oxygen shortage, which can occur during flood-irrigation with poor drainage, may thus increase the susceptibility of salt-sensitive species to salinity.

For many species, acclimation to oxygen-deficient environments involves the emergence of new, adventitious roots from the base of the stem. Gas-filled cavities in the roots are continuous with those in the stem and leaves and allow internal movement of oxygen to the root tip, permitting growth into anaerobic soil (Jackson & Drew, 1984). Cavity formation in the roots (aerenchyma) begins close to the root tip and requires the loss of a large proportion of the root cortex. In maize this does not inhibit ion transport to the xylem (see earlier section). Transport across the cortex might take place symplastically in the remaining radial files of intact cells that bridge the cortical lacunae, but the rate would be much greater than in comparable files of cells in ordinary roots. It would be interesting to know whether secondary formation of plasmodesmata helps augment symplastic transport in this situation.

There are no reports of greater or lesser suberization in roots experiencing poorly-oxygenated environments. In some wetland species (e.g. *Oryza sativa*) the hypodermis becomes suberized close to the root tip, but this is not induced by poor aeration because it also occurs in upland rice varieties. The suberized layer together with the adjacent sclerenchyma probably function in oxygen-deficient media to conserve internal oxygen and restrict the apoplastic entry of potential soil toxins. Restricting the outward radial leakage of oxygen to the environment would divert more to the apical zone for essential respiratory metabolism associated with cell division and growth. Oxygen leakage at the unsuberized tip oxidizes the rhizosphere and de-toxifies some reduced soil components (Armstrong, 1979).

Other edaphic factors

Changes in the physical environment of the root are almost invariably accompanied by alterations to chemical or biotic factors. It is relevant here to mention briefly some of the consequences for root function.

Salinity is an important chemical factor that is often associated both with arid root environments and, where arid climate combines with poor soil drainage, with high water-tables and oxygen shortage. Exclusion of Cl^- by root plasma membranes seems to be linked to salt-induced changes in membrane sterol composition in roots of *Citrus* species (Douglas & Walker, 1984) Exclusion of salt from roots may also be assisted by early formation of structures that block apoplastic pathways across the root. Both *Citrus reticulata* and *Citrus medica* grown in nutrient solution (Walker *et al.*, 1984) deposit suberized lamellae and lignin in hypodermal cell walls (Fig. 5). These block the plasmodesmata (Fig. 6) and the protoplast degenerates, with the exception of unsuberized thin-walled 'passage' cells, every 2-10 cells along the root perimeter. In the endodermis too, suberization (State II) beginning 10-50 mm from the root tip was observed to block plasmodesmata and there was cell death except in unsuberized passage cells located opposite the protoxylem poles. When exposed to saline solution, suberization of hypodermis and endodermis occurred closer to the root tip, probably because of a slowing of root extension. Thus in *Citrus*, it seems probable that transport of ions and water is mainly symplastic, restricted to the passage cells.

Reinforcement of an exclusively symplastic pathway for transport to the xylem, perhaps as a mechanism for controlling or excluding salt in saline environments, is also found in roots of a salt-excluding halophyte *Puccinellia peisonis*, a grass tolerant of poorly-aerated, brackish water (Stelzer & Läuchli, 1977). The endodermis forms two discrete layers, with State II cells occurring as close as 3.5-5.0 mm to the root tip. Passage cells in the biseriate endodermis opposite the xylem poles lack suberin lamellae but develop thickened Casparian bands. Passage cells persist as far as 100-150 mm behind the root tip, but plasmodesmatal continuity is maintained even in the thickened tertiary wall of the other endodermal cells. Measurements of ion distribution in frozen, hydrated sections were made by electron

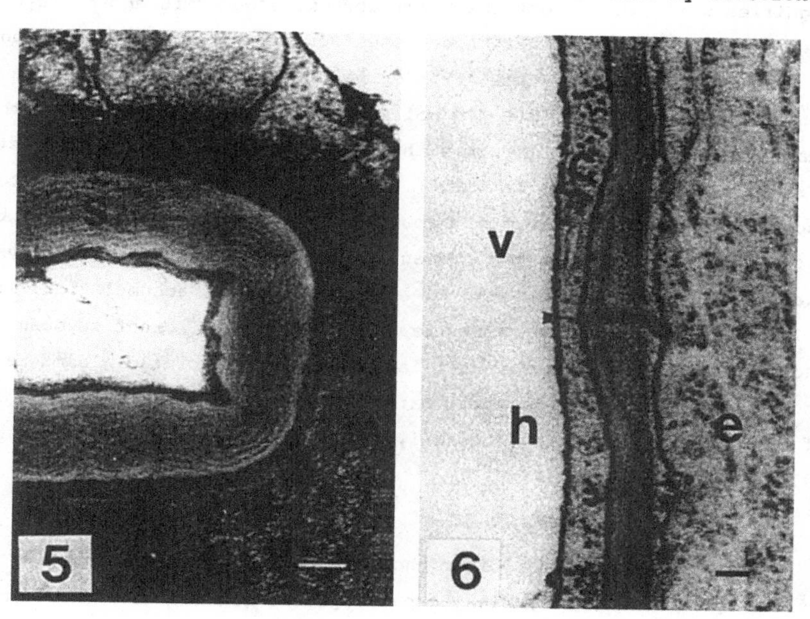

Figs. 5 & 6: Ultrastructure of hypodermis in roots of *Citrus reticulata*. TEM, transverse section. Fig. 5: zone 48-50 mm from root tip of plant treated with 100 mol m^{-3} NaCl showing suberin lamellae (s). bar = 0.1 µm. Fig. 6: zone 4-6 mm from root tip (no NaCl) showing plasmodesmata between epidermal cell (e) and hypodermal cell (h) blocked by suberin lamellae in secondary wall formation (arrowhead); c, cytoplasm; v, vacuole. (from Walker *et al.*, 1984; with permission from Oxford University Press.)

probe X-ray microanalysis (Stelzer & Läuchli, 1978) after roots had been exposed to 200 mol m^{-3} NaCl. They demonstrated a declining Na:K ratio across the root from epidermis to stele as a result of Na 'exclusion' and K accumulation; control over ion transport being augmented perhaps by the double endodermis. The biseriate endodermis does not form in response to environmental stimuli, but is constitutive. However, in various subspecies of *Festuca rubra*, both the glycophyte ssp *fallax* and the halophyte ssp *litoralis* develop thicker endodermal walls in response to greater concentrations of NaCl, those of the halophyte being thicker (Baumeister & Merten, 1981). Under natural conditions, the older roots of the halophyte were characterized by massively thickened lignified walls throughout the inner cortex, endodermis and pericycle, so that it is doubtful whether these zones would function in absorption.

In the halophyte *Atriplex hasta*, which tends to exclude Cl^-, high concentrations of NaCl (100-400 mol m^{-3}) induce wall ingrowths in the outer tangential wall of the epidermis that resemble transfer cell ultrastructure (Kramer, Anderson & Preston, 1978). The induction is not specific to NaCl and seems to occur indirectly through iron deficiency. Electron probe X-ray microanalysis indicated that augmentation of the plasma membrane system in these cells is associated with Cl^- exclusion and K^+ accumulation. Transfer cells are also induced by salt treatments in salt-tolerant soybean varieties and in *Phaseolus coccineus*. Wall ingrowths and associated plasma membrane occur specifically in the xylem parenchyma and may be concerned in resorption of Na^+ from the xylem sap (Läuchli, Kramer & Stelzer, 1974; Kramer *et al*., 1977).

CONCLUSION

There is wide divergence of opinion as to the relative importance of the apoplastic and symplastic pathways for ion and water movement across the root. Entry into the symplast may begin at the epidermis, or at any point within the cortex up to the endodermis. Perhaps, depending on ion concentration, transpiration rate, and root structure, either extreme may apply. The possible pathways are greatly influenced by the location and timing of secondary cell wall developments that are thought to block the apoplastic route, leaving only the symplast for movement across the root. The main barriers to the apoplast are at the endodermis in the form of the Casparian band and suberin lamellae, and similar structures sometimes arise in the hypodermis, although the function there is less well characterized.

The physical environment can modify the timing of development so that precocious suberization or senescence of cells take place. Various functions can be ascribed to these responses. Suberization of the endodermis and hypodermis probably function mainly to protect cells from desiccation by restricting water loss from roots to dry environments, or block passive movement of potentially injurious solutes along apoplastic pathways, or conserve internal oxygen in anaerobic media. Suberized layers also confer a degree of resistance to the penetration of fungal hyphae (Heale, Dodd & Gahan, 1978; Perry & Evert, 1983; Bishop & Cooper, 1983). Some of these adaptations will be important to plant survival of unfavourable environments and account for the tolerance of some species to environmental stress. Other responses are less easily explained: Mg^{2+} deficiency induces increased suberization of the hypodermis and endodermis in maize roots (Pozuelo *et al.*, 1984) and shortage of NO_3^- or NH_4^+ or P_i lead to aerenchyma formation in well-aerated conditions (Konings & Verschuren, 1980; Drew & Saker, 1983).

Transport of ions and water in root tissues are also influenced by the properties of cell membranes and the plasmodesmata that are essential to the symplasmic transport pathway. The influence of the physical environment on these features of cellular organization, apart from temperature adaptation, has received less attention. These aspects, together with a broader knowledge of structure/function relations in roots and their relevance to stress physiology and plant genetic improvement, remain major challenges for the future.

ACKNOWLEDGEMENTS

I am grateful to D.T. Clarkson, P. Olesen, A.W. Robards and R.R. Walker and the CSIRO Division of Horticultural Research for kindly supplying micrographs; and to Jean Nash for typing the manuscript.

REFERENCES

Armstrong, W. (1979). Aeration in higher plants. *Advances in Botanical Research*, 7, 225-331.

Baumeister, W. & Merten, A. (1981). Growth and root anatomy of two subspecies of *Festuca rubra* L. in response to NaCl salinization of the culture solution. *Angewandte Botanik* 55, 401-8.

Bewley, J.D. & Krochko, J.E. 1982. Desiccation-tolerance. In *Physiological Plant Ecology II. Water Relations and Carbon Assimilation. Encyclopedia of Plant Physiology*, vol. 12B, ed. O.L. Lange, P.S. Nobel, C.B. Osmond & H. Ziegler, pp. 325-78. Berlin: Springer-Verlag.

Bishop, C.D. & Cooper, R.M. (1983). An ultrastructural study of root invasion in three vascular wilt diseases. *Physiological Plant Pathology*, 22, 15-27.

Bowling, D.J. (1981). Release of ions to the xylem in roots. *Physiologia Plantarum*, 53, 392-7.

Brouwer, R. & Hoogland, A. (1964). Responses of bean plants to root temperatures. 2. Anatomical aspects. *Jaarboek Instituut voor Biologisch en Scheikundig Onderzoek van Landbouwgewassen, Wageningen* 1964, pp. 23-31.

Cheeseman, J.M. & Hanson, J.B. (1979). Energy-linked potassium influx as related to cell potential in corn roots. *Plant Physiology*, 64, 842-5.

Clark, L.H. & Harris, W.H. (1981). Observations on the root anatomy of rice (*Oryza sativa* L.). *American Journal of Botany*, 68, 154-61.

Clarkson, D.T. (1974). *Ion Transport and Cell Structure in Plants*. London: McGraw Hill.

Clarkson, D.T. (1976). The influence of temperature on the exudation of xylem sap from detached root systems of rye (*Secale cereale*) and barley (*Hordeum vulgare*). *Planta*, 132, 297-304.

Clarkson, D.T. (1984a). Calcium transport between tissues and its distribution in the plant. *Plant, Cell and Environment*, 7, 449-56.

Clarkson, D.T. (1984b). Permeability of hypodermal and epidermal layers in roots of maize. *Agricultural Research Council Letcombe Laboratory Annual Report* 1983, pp. 74-5.

Clarkson, D.T., Hall, K.C. & Roberts, J.K.M. (1980). Phospholipid composition and fatty acid desaturation in the roots of rye during acclimation of low temperature. Positional analysis of fatty acids. *Planta*, 149, 464-71.

Clarkson, D.T., Robards, A.W., Sanderson, J. & Peterson, C.A. (1978). Permeability studies on epidermal-hypodermal sleeves isolated from roots of *Allium cepa* (onion). *Canadian Journal of Botany*, 56, 1526-32.

Clarkson, D.T., Sanderson, J. & Russell, R.S. (1968). Ion uptake and root age. *Nature, London*, 220, 805-6.

Clarkson, D.T., Williams, L. & Hanson, J.B. (1984). Perfusion of onion root xylem vessels: a method and some evidence of control of the pH of the xylem sap. *Planta*, 162, 361-9.

Cole, P.J. & Alston, A.M. (1974). Effect of transient dehydration on absorption of chloride by wheat roots. *Plant and Soil*, 40, 243-7.

de Boer, A.H., Prins, H.B.A. & Zanstra, P.E. (1983). Biphasic composition of trans-root potential in roots of *Plantago* species: Involvement of spatially separated electrogenic pumps. *Planta*, 157, 259-66.

Dell, B. & Wallace, I.M. (1981). Surface root system of *Eucalyptus marginate* Sm.: Anatomy of non-mycorrhizal roots. *Australian Journal of Botany*, 29, 565-77.

Demarty, M., Morvan, C. & Thellier, M. (1984). Calcium and the cell wall. *Plant, Cell and Environment*, 7, 441-8.

Drake, G.A., Carr, D.J. & Anderson, W.P. (1978). Plasmolysis, plasmodesmata, and the electrical coupling of oat coleoptile cells. *Journal of Experimental Botany*, 29, 1205-14.

Douglas, T.J. & Walker, R.R. (1984). Phospholipids, free sterols and adenosine triphosphatase of plasma membrane-enriched preparations from roots of citrus genotypes differing in chloride exclusion ability. *Physiologia Plantarum*, 62, 51-8.

Drew, M.C. (1979). Root development and activities. In *Arid-land Ecosystems: Structure, Functioning and Management*, vol 1, ed. R.A. Perry & D.W. Goodall, pp. 573-606. Cambridge: Cambridge University Press.

Drew, M.C., Chamel, A., Garrec, J.P. & Fourcy, A. (1980). Cortical air spaces (aerenchyma) in roots of corn subjected to oxygen stress. Structure and influence on uptake and translocation of ^{86}rubidium ions. *Plant Physiology*, 65, 506-11.
Drew, M.C. & Dikumwin, E. (1985). Sodium exclusion from the shoots by roots of *Zea mays* (cv LG11) and its breakdown with oxygen deficiency. *Journal of Experimental Botany*, 36, 55-62.
Drew, M.C. & Läuchli, A. (1985). Oxygen-dependent exclusion of sodium ions from shoots by roots of *Zea mays* (cv Pioneer 3906) in relation to salinity damage. *Plant Physiology* 79, 171-6.
Drew, M.C. & Saker, L.R. (1983). Induction of aerenchyma formation by nutrient deficiency in well aerated maize roots. *Agricultural Research Council Letcombe Laboratory Annual Report* 1982, pp.41-2.
Drew, M.C. & Saker, L.R. (1986). Ion transport to the xylem in aerenchymatous roots of *Zea mays* L. *Journal of Experimental Botany*, 37, 22-33.
Dunlop, J. (1982). Membrane potentials in the xylem of roots in intact plants. *Journal of Experimental Botany*, 33, 910-8.
Erwee, M.G. & Goodwin, P.B. (1983). Characterization of the *Egeria densa* Planch. leaf symplast. Inhibition of the intercellular movement of fluorescent probes by group II ions. *Planta*, 158, 320-8.
Everard, J.D. & Drew, M.C. (1983). Effects of oxygen deficiency on root hydraulic conductivity and shoot water relations. *Agricultural Research Council Letcombe Laboratory Annual Report* 1982, pp.39-40.
Ferguson, I.B. & Clarkson, D.T. (1975). Ion transport and endodermal suberization in the roots of *Zea mays*. *New Phytologist*, 75, 69-79.
Ferguson, I.B. & Clarkson, D.T. (1976). Ion uptake in relation to the development of a root hypodermis. *New Phytologist*, 77, 11-4.
Guiaquinta, R.T., Lin, W., Sadler, N.L., & Franceschi, V.R. (1983). Pathway of phloem unloading of sucrose in corn roots. *Plant Physiology*, 72, 362-7.
Goodwin, P.B. (1983). Molecular size limit for movement in the symplast of the *Elodea* leaf. *Planta*, 157, 124-30.
Grunwaldt, G., Ehwald, R., Pietzsch, W. & Göring, H. (1979). A special role of the rhizodermis in nutrient uptake by plant roots. *Biochemie und Physiologie der Pflanzen*, 174, 831-7.
Gunning, B.E.S. & Robards, A.W. (1976). Plasmodesmata and symplastic transport. In *Transport and Transfer Processes in Plants*, ed. I.E. Wardlaw & J.B. Passioura, pp. 15-41. New York: Academic Press.
Hanson, J.B. (1978). Application of the chemiosmotic hypothesis in ion transport across the root. *Plant Physiology*, 62, 402-5.
Harrison-Murray, R.S. & Clarkson, D.T. (1973). Relationships between structural development and the absorption of ions by the root system of *Cucurbita pepo*. *Planta*, 114, 1-16.
Heale, J.B., Dodd, K.S. & Gahan, P.B. (1978). The induced resistance response of carrot root slices to heat-killed conidia and cell-free germination fluid of *Botrytis cinerea* Pes. ex Pers. 1. The possible role of cell death. *Annals of Botany*, 49, 847-57.
Henry, C.M. & Deacon, J.W. (1981). Natural (non-pathogenic) death of the cortex of wheat and barley seminal roots, as evidenced by nuclear staining with acridine orange. *Plant and Soil*, 60, 255-74.
Hepler, P.K. (1982). Endoplasmic reticulum in the formation of the cell plate and plasmodesmata. *Protoplasma*, 111, 121-33.

Iren, F. van & Boers, van der Sluijs, P. (1980). Symplasmic and apoplasmic radial ion transport in plant roots. Cortical plasmalemmas lose absorption capacity during differention. *Planta*, 148, 130-7.

Jackson, M.B. & Drew, M.C. (1984). Effects of flooding on growth and metabolism of herbaceous plants. In *Flooding and Plant Growth*, ed. T.T. Kozlowski, pp. 47-128. New York: Academic Press.

Jones, H., Tomos, A.D., Leigh, R.A. & Wyn Jones, R.G. (1983). Water-relations parameters of epidermal and cortical cells in the primary root of *Triticum aestivum* L. *Planta*, 158, 230-6.

Kochian, L.V. & Lucas, W.J. (1983). Potassium transport in corn roots. 2. The significance of the root periphery. *Plant Physiology*, 73, 208-15.

Kolattukudy, P.E. (1980). Cutin, suberin, and waxes. In *The Biochemistry of Plants: A Comprehensive Treatise*, vol. 4, *Lipids: Structure and Function*, ed. P.K. Stumpf, pp. 571-645. New York: Academic Press.

Konings, H. & Verschuren, G. (1980). Formation of aerenchyma in roots of *Zea mays* in aerated solutions, and its relation to nutrient supply. *Physiologia Plantarum*, 49, 265-70.

Kramer, D., Anderson, W.P. & Preston, J. (1978). Transfer cells in the root epidermis of *Atriplex hastata* L. as a response to salinity: a comparative cytological and X-ray microprobe investigation. *Australian Journal of Plant Physiology*, 5, 739-47.

Kramer, D., Läuchli, A., Yeo, A.R. & Gullasch, J. (1977). Transfer cells in roots of *Phaseolus coccineus*: Ultrastructure and possible function in exclusion of sodium from the shoot. *Annals of Botany*, 41, 1031-40.

Kramer, P.J. (1983). *Water Relations of Plants*. New York: Academic Press.

Kuo, J., McComb, A.J. & Cambridge, M.L. (1982). Ultrastructure of the seagrass rhizosphere. *New Phytologist*, 89, 139-43.

Kurkova, E.B. (1981). Distribution of plasmodesmata in root epidermis. In *Structure and Function of Plant Roots*, ed. R. Brouwer, O. Gasparikova, J. Kolek & B.C. Loughman, pp. 107-9. The Hague: Martinus Nijhof/Dr W. Junk.

Läuchli, A. (1976a). Apoplasmic transport in tissues. In *Transport in plants. Encyclopedia of Plant Physiology*, vol. 2B, ed. U. Lüttge & M.G. Pitman, pp. 3-34. Berlin: Springer-Verlag.

Läuchli, A. (1976b). Symplasmic transport and ion release to the xylem. In *Transport and Transfer Processes in Plants*, ed. I.E. Wardlaw & J.B. Passioura, pp. 101-12. New York: Academic Press.

Läuchli, A. (1984). Salt exclusion: an adaptation of legumes for crops and pastures under saline conditions. In *Salinity Tolerance in Plants. Strategies for Crop Improvement*, ed. R.C. Staples & G.H. Toenniessen, pp. 171-87. New York: John Wiley & Sons.

Läuchli, A., Kramer, D., Pitman, M.G. & Lüttge, U. (1974a). Ultrastructure of xylem parenchyma cells of barley roots in relation to ion transport to the xylem. *Planta*, 119, 85-99.

Läuchli, A., Kramer, D., & Stelzer, R. (1974b). Ultrastructure and ion localization in xylem parenchyma cells of roots. In *Membrane Transport in Plants*, ed. U. Zimmerman & J. Dainty, pp. 363-71. Berlin: Springer-Verlag.

Leshem, B. (1970). Resting roots of *Pinus halepensis*: Structure, function, and reaction to water stress. *Botanical Gazette*, 131, 99-104.

Lyons, J.M., Raison, J.K. & Steponkus, P.L. (1979). The plant membrane in response to low temperature: an overview. In *Low Temperature Stress in Crop Plants*, ed. J.M. Lyons, D. Graham & J.K. Raison, pp. 1-24. New York: Academic Press.

Mees, G.C. & Weatherley, P.E. (1957). The mechanism of water absorption by roots. 2. The role of hydrostatic pressure gradients across the cortex. *Proceedings of the Royal Society Series B*, 141, 381-91.
Newman, E.I. (1976). Water movement through root systems. *Philosophical Transactions of the Royal Society London B*, 273, 463-478.
Nir, I., Klein, S. & Poljakoff-Mayber, A. (1969). Effect of moisture stress on submicroscopic structure of maize roots. *Australian Journal of Biological Sciences*, 22, 17-33.
Nobel, P.S. & Sanderson, J. (1984). Rectifier-like activities of roots of two desert succulents. *Journal of Experimental Botany*, 35, 727-37.
Okomoto, H., Ichino, K. & Katou, K. (1978). Radial electrogenic activity in the stem of *Vigna sesquipedalis*: involvement of spatially separate pumps. *Plant, Cell and Environment*, 1, 279-84.
Olesen, P. (1978). Studies on the physiological sheats in roots. 1. Ultrastructure of the exodermis in *Hoya carnosa* L. *Protoplasma*, 94, 325-40.
Olesen, P. (1979). The neck constriction in plasmodesmata: evidence for a peripheral sphincter-like structure revealed by fixation with tannic acid. *Planta*, 144, 349-58.
Olesen, P. (1980). A model of a possible sphincter associated with plasmodesmatal neck regions. *European Journal of Cell Biology*, 22, 250.
Osmond, D.L., Wilson, R.F. & Raper, C.D. (1982). Fatty acid composition and nitrate uptake of soybean roots during acclimation to low temperature. *Plant Physiology*, 70, 1689-93.
Overall, R.L. & Gunning, B.E.S. (1982). Intercellular communication in *Azolla* roots: 2. Electrical coupling. *Protoplasma*, 111, 151-60.
Overall, R.L., Wolfe, J. & Gunning, B.E.S. (1982). Intercellular communication in *Azolla* roots: 1. Ultrastructure of plasmodesmata. *Protoplasma*, 111, 134-50.
Parsons, L.R. & Kramer, P.J. (1974). Diurnal cycling in root resistance to water movement. *Physiologia Plantarum*, 30, 19-23.
Perry, J.W. & Evert, R.F. (1983). The effect of colonization by *Verticillium dahliae* on root tips of Russett Burbank potatoes. *Canadian Journal of Botany*, 61, 3422-9.
Peterson, C.A., Emanuel, M.E. & Humphreys, G.B. (1981). Pathways of movement of apoplastic fluorescent dye tracers through the endodermis at the site of secondary root formation in corn (*Zea mays*) and broad bean (*Vicia faba*). *Canadian Journal of Botany*, 59, 618-25.
Peterson, C.A., Emanuel, M.E. & Wilson, C. (1982). Identification of a Casparian band in the hypodermis of onion and corn roots. *Canadian Journal of Botany*, 60, 1529-35.
Peterson, C.A. & Perumalla, C.J. (1984). Development of the hypodermal Casparian band in corn and onion roots. *Journal of Experimental Botany*, 35, 51-7.
Peterson, C.A., Peterson, R.L. & Robards, A.W. (1978). A correlated histochemical and ultrastructural study of the epidermis and hypodermis of onion roots. *Protoplasma*, 96, 1-21.
Pitman, M.G. (1982). Transport across plant roots. *Quarterly Reviews of Biophysics*, 15, 481-554.
Powell, D.B.B. (1978). Regulation of plant water potential by membranes of the endodermis in young roots. *Plant, Cell and Environment*, 1, 69-76.

Pozuelo, J.M., Espie, K.E. & Kolattukudy, P.E. (1984). Magnesium deficiency results in increased suberization in endodermis and hypodermis of corn roots. *Plant Physiology*, 74, 256-60.

Rao, K.P. & Rains, D.W. (1976). Nitrate absorption by barley. l. Kinetics and energetics. *Plant Physiology*, 57, 55-8.

Ritchie, R.J. & Larkum, A.W.D. (1982). Ion exchange fluxes of the cell walls of *Enteromorpha intestinalis* (L.) Link (Ulvales, Chlorophyta). *Journal of Experimental Botany*, 33, 140-53.

Robards, A.W. (1976). Plasmodesmata in higher plants. In *Intercellular Communication in Plants: Studies on Plasmodesmata*, ed. B.E.S. Gunning & A.W. Robards, pp. 15-57, Berlin: Springer-Verlag.

Robards, A.W. & Clarkson, D.T. (1984). Effects of chilling temperatures of root cell membranes as viewed by freeze-fracture electron microscopy. *Protoplasma*, 122, 75-85.

Robards, A.W., Clarkson, D.T. & Sanderson, J. (1979). Structure and permeability of the epidermal/hypodermal layers of the sand sedge (*Carex arenaria*, L.). *Protoplasma*, 101, 331-47.

Robards, A.W., Jackson, S.M., Clarkson, D.T. & Sanderson, J. (1973). The structure of barley roots in relation to the transport of ions into the stele. *Protoplasma*, 77, 291-311.

Robards, A.W. & Robb, M.E. (1974). The entry of ions and molecules into roots: an investigation using electron-opaque tracers. *Planta*, 120, 1-12.

Roberts, J.K.M., Callis, J., Jardetsky, O., Walbot, V. & Freeling, M. (1984). Cytoplasmic acidosis as a determinant of flooding intolerance in plants. *Proceedings of the National Academy of Sciences, U.S.A*, 81, 6029-33.

Saglio, P.H., Raymond, P. & Pradet, A. (1980). Metabolic activity and energy charge of excised maize root tips under anoxia. Control by soluble sugars. *Plant Physiology*, 66, 1053-57.

Sanderson, J. (1983a). Water uptake by different regions of the barley root. Pathways of radial flow in relation to development of the endodermis. *Journal of Experimental Botany*, 34, 240-53.

Sanderson, J. (1983b). Effect of transpiration of translocation of calcium and phosphate from different regions of the barley root. *Agricultural Research Council Letcombe Laboratory Annual Report* 1982, pp. 72-3.

Schönherr, J. & Ziegler, H. (1980). Water permeability of *Betula* periderm. *Planta*, 147, 345-54.

Seagull, R.W. (1983). Differences in the frequency and disposition of plasmodesmata resulting from root cell elongation. *Planta*, 159, 497-504.

Sentenac, H. & Grignon, C. (1981). A model for predicting ionic equilibrium concentrations in cell walls. *Plant Physiology*, 68, 415-9.

Shone, M.G.T. & Flood, A.V. (1983). Effects of periods of localized water stress on subsequent nutrient uptake by barley roots and their adaptation by osmotic adjustment. *New Phytologist*, 94, 561-72.

Spanswick, R.M. (1976). Symplasmic transport in tissues. In *Transport in Plants. Encyclopedia of Plant Physiology*, vol. 2B, ed. U. Lüttge & M.G. Pitman, pp. 35-53. Berlin: Springer-Verlag.

Stelzer, R. & Läuchli, A. (1977). Salz-und Uberflutungstoleranz von Puccinellia peisonis. 2. Strukturelle differenzierung der wurzel in beziehung zur function. *Zeischrift für Pflanzenphysiologie*, 84, 95-108.

Stelzer, R. & Läuchli, A. (1978). Salt-and flooding-tolerance of *Puccinellia peisonis*. 3. Distribution and localization of ions in the plant. *Zeischrift für Pflanzenphysiologie*, 88, 437-48.

Stephens, J.S. (1981). Effects of Temperature on Hydraulic Conductivity of the Roots of *Zea mays*. England: Ph.D. Thesis, University of Reading.
Stephens, J.S. & Clarkson, D.T. (1981). Water uptake and hydraulic conductivity in various zones of maize roots. *Agricultural Research Council Letcombe Laboratory Annual Report* 1980, pp. 62-3.
Steudle, E. & Jeschke, W.D. (1983). Water transport in barley roots. Measurements of root pressure and hydraulic conductivity of roots in parallel with turgor and hydraulic conductivity of root cells. *Planta*, 158, 237-48.
Tippett, J.T. & O'Brien, T.P. (1976). The structure of eucalypt roots. *Australian Journal of Botany*, 24, 619-22.
Trought, M.C.T. & Drew, M.C. (1980). The development of waterlogging damage in young wheat plants in anaerobic solution culture. *Journal of Experimental Botany*, 31, 1573-85.
Tyree, M.T. (1970). The symplast concept. A general theory of symplastic transport according to the theory of irreversible processes. *Journal of Theoretical Biology*, 26, 181-214.
Uemura, M. & Yoshida, S. (1984). Involvement of plasma membrane alterations in cold acclimation of winter rye seedlings (*Secale cereale* L. cv Puma). *Plant Physiology*, 75, 818-26.
Vakhmistrov, D.B. (1981). Specialization of root tissues in ion transport. In *Structure and Function of Plant Roots*, ed. R. Brouwer, O. Gasparikova, J. Kolek & B.C. Loughman, pp. 203-8. The Hague: Martinus Nijhof/Dr W. Junk.
Vakhmistrov, D.B. & Kurkova, E.B. (1979). Symplastic connections in rhizodermis of *Trianea bogotensis*. *Fiziologiya Rastenii*, 26, 943-52.
Vakhmistrov, D.B., Mel'nikov, P.V. & Vorob'ev, L.N. (1974). Differences in absorption of potassium by root hairs and hairless cells of the root epidermis in *Trianea bogotensis*. *Fiziologiya Rastenii*, 21, 554-62.
Vogt, E., Schönherr, J. & Schmidt, H.W. (1983). Water permeability of periderm membranes isolated enzymatically from potato tubers (*Solanum tuberosum* L.). *Planta*, 158, 294-301.
Walker, N.A. & Pitman, M.G. (1976). Measurement of fluxes across membranes. In *Transport in Plants. Encyclopedia of Plant Physiology*, vol. 2B, ed. U. Lüttge & M.G. Pitman, pp. 93-126. Berlin: Springer-Verlag.
Walker, R.R., Sedgley, M., Blesing, M.A. & Douglas, T.J. (1984). Anatomy, ultrastructure and assimilate concentrations of roots of citrus genotypes differing in ability for salt exclusion. *Journal of Experimental Botany*, 35, 1481-94.
Weatherley, P.E. (1982). Water uptake and flow in roots. In *Physiological Plant Ecology II. Water Relations and Carbon Assimilation. Encyclopedia of Plant Physiology*, vol. 12B, ed. O.L. Lange, P.S. Nobel, C.B. Osmond & H. Ziegler, pp. 79-109. Berlin: Springer-Verlag.
Wilcox, H.E. (1968). Morphological studies of the root of red pine, *Pinus resinosa*. 1. Growth characteristics and pattern of branching. *American Journal of Botany*, 55, 247-54.
Wilson, A.J. & Robards, A.W. (1978). The ultrastructural development of mechanically impeded barley roots. Effects on the endodermis and pericycle. *Protoplasma*, 95, 255-65.
Wullstein, L.H. & Pratt, S.A. (1981). Scanning electron microscopy of rhizoheaths of *Oryzopsis hymenoides*. American Journal of Botany, 68, 408-19.

ORIGIN, BRANCHING AND DISTRIBUTION OF ROOT SYSTEMS

Betty Klepper

INTRODUCTION

To most people, root systems are the large, well-branched, deep networks of roots revealed by excavations under mature plants. However, these large networks have arisen from individual root growth and branching patterns characteristic of a species, from root/shoot interactions and from several abiotic soil factors. These abiotic factors include distributions in the soil profile of temperature, water content, oxygen and strength and the presence of inhibitory soil factors such as cemented layers, high concentrations of such chemicals as aluminium, and the presence of low osmotic potentials.

This paper reviews some of the plant and soil factors important in determining the distribution of root systems with primary emphasis on annual crop plants and only limited concern for perennial plants.

ORIGIN OF ROOTS

Although roots can originate from a wide variety of tissues, they generally arise from the primary root or its branches, from stem nodes in stolons, bulbs, and grasses, or from the hypocotyl of certain dicotyledonous plants (dicots). More than one type of root can occur in one species. For example, clover (***Trifolium repens*** **L.**) has a taproot system as well as an extensive system of roots produced at nodes on stolons (Ueno, 1982). Production of primordia of either lateral or adventitious roots appears to coincide with the presence of xylem vessels near parenchymatous cells capable of dedifferentiation and undoubtedly involves complex hormonal responses (Friedman, Altman & Zamski, 1979).

Dicotyledons

Most dicot root systems originate from the primary root and form a tap-rooted system. Taproots are characterized by having one strongly orthogeotropic (vertical) central root with subordinate laterals. Taproots range from strongly taprooted species such as sugar beet (*Beta vulgaris* L.) with a fleshy storage root, to cotton (*Gossypium hirsutum* L.) with a woody taproot, to fibrous systems which are a modified form of the tap-rooted system. For example, modified systems are commonly seen in such legumes as bean (*Phaseolus vulgaris* L.) or pea (*Pisum sativum* L.) and occur when the radicle atrophies shortly after emergence and a flush of laterals takes over soil exploration with no lateral being especially dominant. The primary difference between dicots with strongly taprooted systems and those with diffuse systems appears to arise from the degree of apical dominance expressed, especially upon the death of the taproot. If cotton or carrot taproots are injured or killed, one lateral will grow faster than the others and will re-establish a single dominant root when the injury occurs in the upper part of the profile. When a bean root is injured, several laterals take over the downward penetration of the soil. For all dicots, however, the relative dominance of the taproot generally wanes with depth and even strongly tap-rooted species eventually behave like diffusely rooted species (Riedacker, *et al.*, 1982).

Most dicot roots are capable of secondary growth and develop complex vascular relationships among parent roots and their branches. For example, lateral roots develop secondary xylem which is continuous with parent root xylem on the upper side (Klepper, 1983) but not usually on the lower side, which often appears to have a layer or pad of "corky" material separating the xylem of the lateral root from that of the parent root (Klepper, unpublished observations).

Each lateral root probably serves a small section of the shoot and certain areas of leaf are "targeted" for the delivery of materials from particular roots. This is illustrated by the colour patterns of wheat leaves infected with *Cephalosporium* stripe. The pathogen enters the plant through individual damaged root axes and travels to the shoot only in the xylem of those leaf veins supplied by the damaged roots. The characteristic yellow striping of the leaves is caused by the chlorotic tissue affected by pathogens in those leaf veins supplied by the damaged roots in contrast to the normal tissue supplied by the rest of the root system. However, lateral exchange and mixing of water and minerals in xylem readily occurs and evens out any localized deficiencies of supply.

Monocotyledons

Monocot roots (except for the radical) usually originate from stem nodes although some arise from internodes (Martin & Harris, 1976). In the grass family, these nodal roots consist of two root systems: the seminal system which arises from root primordia associated with nodes in the embryo, and the crown root system which comes from primordia associated with nodes in the crown (Klepper, Belford & Rickman, 1984). This discussion emphasizes wheat, but other grasses can be analyzed similarly (Hoshikawa, 1969).

The seminal root system of wheat is made up of the primary root and up to five roots associated with the two nodes (scutellar and epiblast) in the seed. The seminal root system originates at the depth of planting. Under some conditions, adventitious roots associated with the coleoptilar node in wheat also originate at this depth (Peterson, Klepper & Rickman, 1982). The crown root system develops at the depth of the crown and arises from successively higher nodes with the passage of time (Klepper *et al.*, 1984). The number and diameter of these roots increases with higher node number for corn (Yamazaki & Kaeriyama, 1982), wheat (R.K. Belford, unpublished data), and millet (Gregory, 1983). Generally there are 2-4 per node.

A general scheme for identifying and naming nodal axes of cereals is shown in Fig. 1. Each root axis is named after the node with which it is associated and is further defined by describing the direction of growth with respect to the leaf borne at the same node. For example, a root which arises from node ($\underline{n}$+1) and growing in the quadrant of the node which is toward the leaf at that node would be called the ($\underline{n}$+1)X root. When more than one root arises from a quadrant, the less-developed roots are given primed superscripts. For example five roots at the 5th node might be 5A, 5B, 5X, 5Y and 5Y'. In this case, 5Y' would be the younger, less-developed of the two roots arising from the side of the node opposite to the midrib of the leaf attached at that node. Fig. 2 shows an abnormal wheat crown which has developed internodes in the crown. The figure clearly shows that the crown roots are associated with these nodes.

BRANCHING OF ROOTS

Each root segment formed in a root system passes through a maturation sequence which includes initiation of lateral primordia 1 to 2 cm proximal to the tip (Hackett & Stewart, 1969; MacLeod & Thompson, 1979), production of root hairs in the most recently elongated tissue, and emergence of lateral roots in the zone immediately proximal to the root hair zone. The distance between the root tip and the point of appearance of young lateral roots depends on the species and the growing conditions. Fig. 3 shows the effect of temperature on this distance for roots of bean (*Phaseolus vulgaris* L.) (Brouwer & Hoagland, 1964). For this species, temperature effects are significant only below 15°C. Table 1 shows features of primary roots of four species grown at room temperature. The distance from the tip of the youngest lateral root ranges from 5.5 to 8.2 cm for the three legumes, but is greater (11.4 cm) for maize (MacLeod & Thompson, 1979) under the conditions of this experiment. In rice, this distance was found to be greater in thick, rapidly-growing than in thin, slowlygrowing axes (Sasaki, Yamazaki & Kawata, 1984).

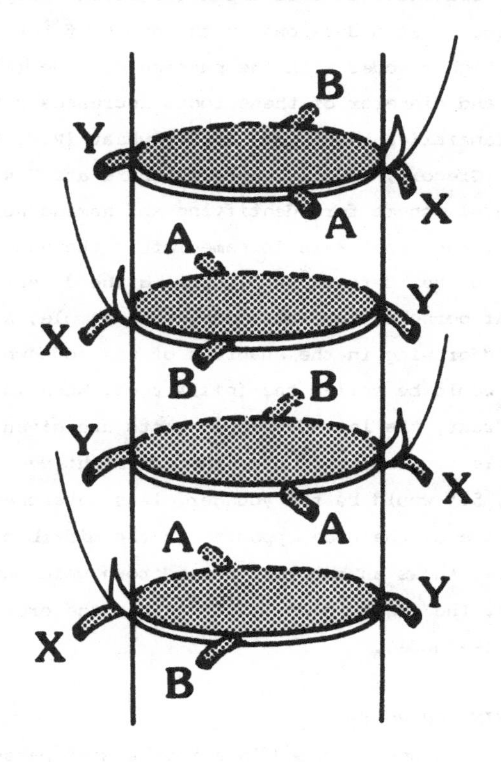

Fig. 1. Diagram of successive nodes on a cereal plant showing a node-based root identification scheme (Modified from Klepper *et al.*, 1984).

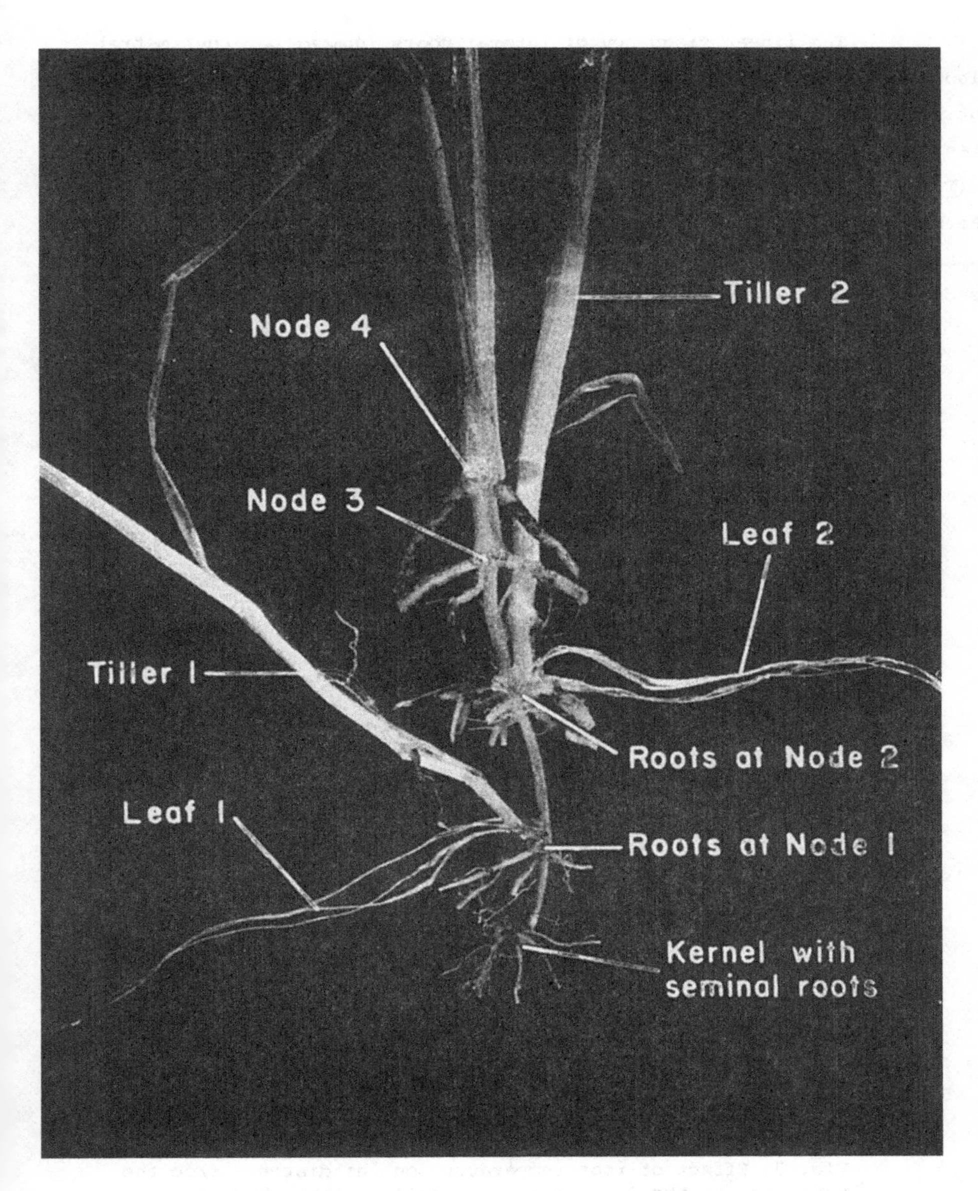

Fig. 2. An abnormal wheat crown with elongated internodes. Individual nodes, tillers, leaf scars and roots are named.

The linear frequency of lateral roots (number per centimetre) also varies with species and with culture conditions. There is an effect of nutrients (Hackett, 1972), especially if there are marked nutrient deficiencies (Tennant, 1976a). Legumes have about 5 roots per centimetre and maize 8 (Table 1), radish 5 to 6 (Blakely *et al.*, 1982), and barley 1.7-3.7 (Hackett & Bartlett, 1971) or 5 to 6 (May, Chapman & Aspinall, 1965). Apparently, there are significant effects of environment, even under optimized conditions in the laboratory, on linear frequency. The number of lateral

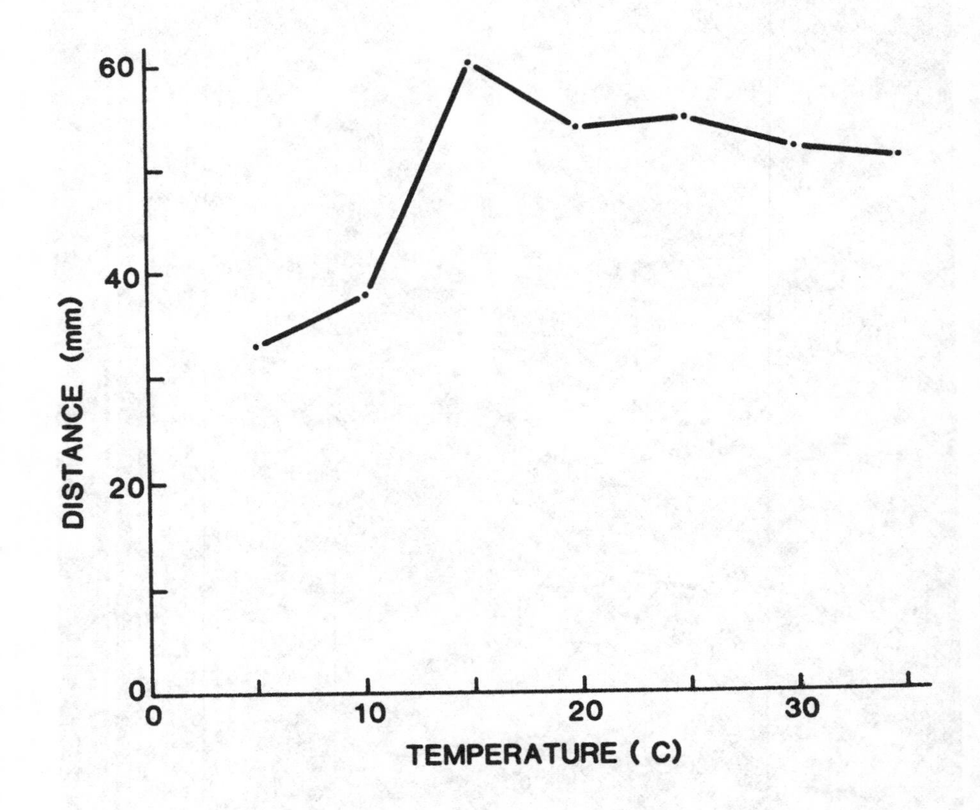

Fig. 3. Effect of root temperature on the distance from the root apex to the first macroscopically visible branch root in beans (Redrawn from Brouwer & Hoagland, 1964).

roots may be correlated with the diameter of the parent axis (Yamazaki, Sasaki & Kawata, 1981; Sasaki, Yamazaki & Kawata, 1983; Yamazaki & Kaeriyama, 1983). Unfortunately, most data are from laboratory experiments concerned only with the earliest roots of seedlings under experimental conditions which are often far-removed from those found in the field.

In a recent field experiment (Belford, *et al.*, 1982), intact root systems of winter wheat were extracted from the field at different times in the growing season. The individual root axes were identified (Klepper *et al.*, 1984) and examined for degree and linear frequency of branching. Those axes which developed branches after spring fertilizer was applied showed an increase in the linear frequency of branching (R.K. Belford, personal communication).

The spatial arrangement of root branches on the parent root appears to be non-random. Lateral roots generally form opposite the xylem poles in dicots and can often be seen to occur in ranks (Charlton, 1983). In a study of five species, Mallory *et al.* (1970) found that the degree of

Table 1. Morphological features of primary root systems on seedlings of four species grown for up to 12 days under laboratory conditions (from MacLeod & Thompson, 1979).

Species	Vicia faba	Pisum Sativum	Zea mays	Phaseolus vulgaris
Distance of primordia from tip (cm)*	2.3	1.6	3.2	2.1
Distance of youngest lateral from tip (cm)	8.2	5.5	11.4	5.6
Primordia formed				
(number/cm)	5.4	3.9	8.4	6.8
(number/day)	11.0	7.8	25.2	12.4
Laterals (number/cm)	6.2	4.7	7.7	6.8
Time from primordia formation to appearance of lateral (days)	3.1	3.1	3.2	2.4

*Visible under dissecting microscope.

regularity is inversely related to the number of protoxylem poles. Riopel (1969) found a non-random dispersed pattern of lateral roots in six monocot species. As Mallory *et al*., (1970) pointed out, these patterns are often difficult to discern because not all the lateral primordia develop into lateral roots and the pattern is presumably set by the initiation of primordia.

In monocots, the unique relationship between stem nodes and crown root axes permits correlations between shoot and root development to be derived (Klepper *et al*., 1984). For example, Fig. 4 shows the degree of branching found on individually identified root axes and the number of leaves on the main stem as a function of phyllochrons after emergence for wheat plants. A phyllochron is the unit of time between equivalent growth stages of successive leaves, *e.g.*, between a visible leaf number of $\underline{n}$ and $\underline{n}+1$. This diagram shows that each identified axis produces branches about 2.5 phyllochrons afrer it first appears. Since each phyllochron requires approximately 100 growing degree days (Rickman, Klepper & Belford, 1985), then the appearance and branching of wheat root axes can also be predicted from cumulative growing degree days. Cumulative growing degree days is calculated by summing the average air temperature in Celsius degrees for each day with all negative means taking a value of zero. A plant sampled 600 growing degree days after emergence would be expected to have 6 main stem leaves, well-branched seminal root axes and a crown root system with branching only on the axes associated with the first foliar node (Fig. 4). Similar information has been developed for rice plants (Yamazaki & Harada, 1982).

DISTRIBUTION OF ROOTS

The main factors influencing the distribution of root systems are genetic differences in species and cultivars, soil textural, physical and chemical properties, and seasonal shifts in distribution of root primordia.

Genetic factors

Most cultivars have been selected on the basis of shoot properties, but where differences in shoot morphology are associated with differences in root systems, there may have been inadvertent selection for root properties (Troughton & Whittington, 1969). For example, certain cultivars of alfalfa

(*Medicago sativa* L.) with taller, heavier shoots also have more root branching (McIntosh & Miller, 1981). Likewise, there is a correlation between root and shoot morphology in white clover (*Trifolium repens* L.) (Caradus, 1981). Derera, Marshall & Balaam, (1969) found varietal differences in root fineness (length per unit root weight) among wheat varieties. Hurd (1968) found some varietal differences in growth rates and rooting patterns of wheat roots, but also showed that the effects of soil type and physical condition were in general greater than were varietal effects. This agrees with results

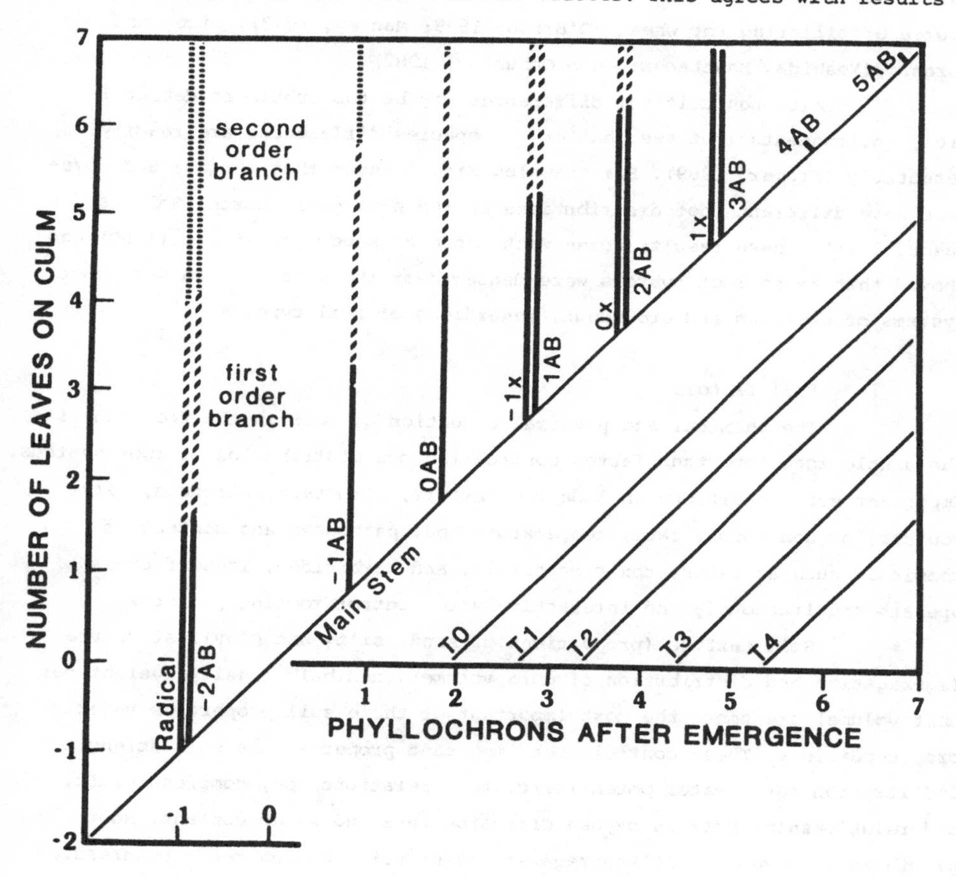

Fig. 4. Relationship between shoot and root development on main stems of wheat (Modified from Klepper *et al.*, 1984).

on alfalfa (McIntosh & Miller, 1981), and may explain some of the contradictions in the literature. For example, cultivars of tall and semi-dwarf wheat have significant differences in rooting density and distribution under field and greenhouse conditions (Köpke, Böhm & Jachmann, 1982) but no significant differences under dryland field conditions (Cholick, Welsh & Cole, 1977). Where studies have separated seminal and nodal root systems, the seminal root system has shown significant genetic variation (O'Brien, 1979) but the size of the crown root system has been related primarily to degree of tillering for wheat (O'Brien, 1979; MacKey, 1973), rice, and sorghum (Yoshida, Bhattacharjee & Cabuslay, 1982).

Although cultivar differences may be too subtle to detect in field environments (but see Chapter 7), species differences are readily detectable (Kramer, 1969). For example, Fig. 5 shows that sorghum and soyabean have different root distributions in the same soil (Burch, Smith & Mason, 1978). These results agree with those of Robertson *et al.* (1980) who showed that maize root systems were denser near the surface than were root systems of soyabean and groundnut, regardless of soil type.

Soil factors

The chemical and physical condition of soil in the root zone is the single most important factor controlling the distribution of root systems. Important soil conditions include the texture, structure, strength, water content, oxygen supply rate, temperature, pH, pathogens and content of chemicals such as salts, toxic materials, and herbicides. These factors operate simultaneously and interactively to control rooting patterns.

Soil texture (proportions of sand, silt, and clay), structure (aggregation and distribution of pore volume), and bulk density (weight per unit volume) are among the most important of these soil properties under crop conditions. These control such important properties as soil strength, infiltration rate, water potential/content relationships, compressibility and relationships between oxygen diffusion rate and water content. Root growth is favoured by well-aggregated, moist soils having ready compressibility, relatively low strength, high fertility, and an optimum balance between air and water in pore spaces. Roots generally proliferate where conditions are good and remain relatively static where conditions are poor. Because soil strength is so often a controlling factor, roots often grow along pre-existing cracks (Whiteley & Dexter, 1982), worm holes (Ehlers *et al.*, 1983; Edwards & Lofty, 1978), or old root channels. Failing these

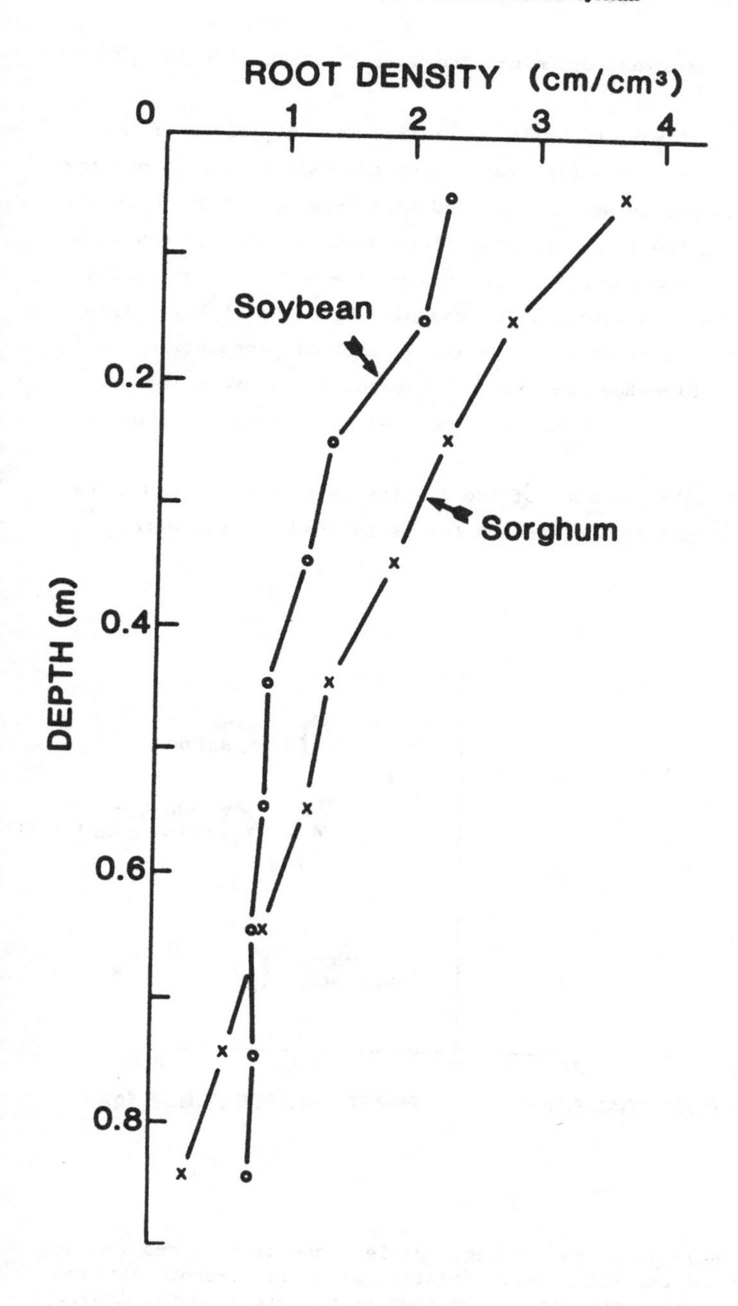

Fig. 5. Distribution of sorghum and soyabean root systems (Redrawn from Burch *et al.*, 1978).

easy routes, they must compress or deform the soil (Whiteley & Dexter, 1984).

In a range of soil types in Australia, Tennant (1976b) found that wheat roots penetrated at faster rates (about 2 cm/day) and to greater depths (1.7m) in a sandy loam and a deep sand than in a grey clay where the rate of penetration was about 25% and the final depth of penetration less than 20% of that in the sandier soils. Generally, however, sand or gravel do not especially favour root growth. For example, Vine, Lal & Payne (1981) found that root lengths of maize were 27% less, depth of penetration 65% less, and seminal root diameter 36% greater in sands and gravels than in forest topsoil. Loamy soils are generally the most favourable for root proliferation.

Fig. 6 illustrates some of the complex relationships found for elongating radicles of pea (*Pisum sativum* L.) as related to soil water

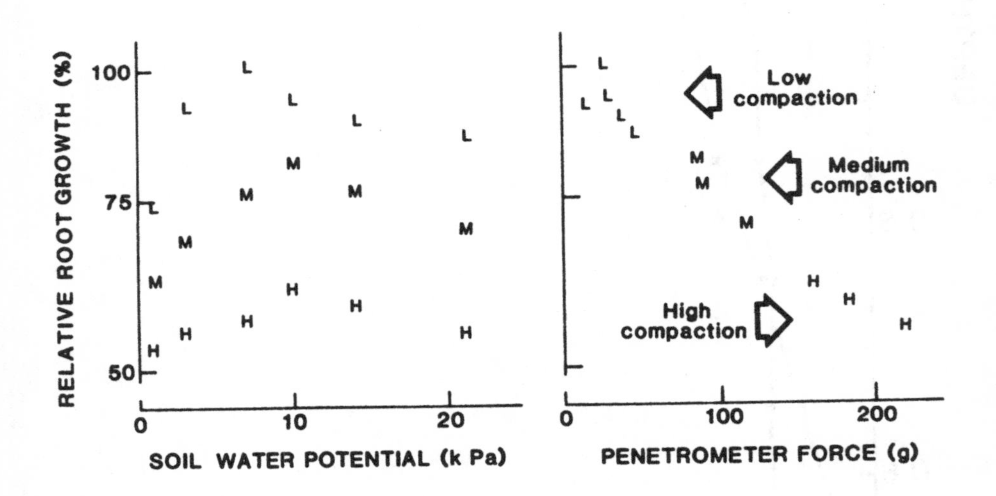

Fig. 6. Relationships between radicle elongation in pea (*Pisum sativum* L.) and soil water potential and soil strength at three levels of soil compaction. (Redrawn from Eavis & Payne, 1969).

potential and soil strength (penetrometer resistance)(Eavis & Payne, 1969). The effects of poor aeration are shown by the decreased root growth in wetter soils at all three bulk densities. The dramatic effects of increasing soil strength on root growth are clearly shown but note that the effect of soil water potential, *per se*, is not very important. These data illustrate the importance of relating root growth to fundamental physical parameters such as strength instead of to bulk density (Taylor & Gardner, 1963).

The driving force for root penetration of soil results from the turgor pressure in elongating root cells as it is restrained by the cell wall and the surrounding soil (Greacen & Oh, 1972). The osmotic potential of root cells is generally about -700 to -1200 kPa (Taylor & Ratliff, 1969), approximately half the osmotic potential commonly found in shoots. The mechanical advantage provided by the wedge-shaped root cap, the presence of localized weak points in soil, osmotic adjustment of root cells, possibly the nutation of growing roots (Ney & Pilet, 1981), and the presence of mucilages at root tips allow roots to penetrate soils with resistances greater than the osmotic potential in root cells. Ehlers *et al.* (1983) found that oat (*Avena sativa* L.) roots ceased growth at 3600 kPa soil penetrometer resistance in tilled soil which had no large pores from old root or worm channels, but at 4600-5100 kPa in untilled soils which did have these pores. Compacted layers in the soil cause a build-up of roots above the layer, a decrease in root length within the layer, and an increase in root diameter within the layer (Shierlaw & Alston, 1984; Boone & Veen, 1982). This response probably results from the proliferation of laterals when the main root is injured (Hackett, 1971).

The growth rate of roots is markedly affected by soil water content through its effect on soil strength and its usually less-important effect on soil water potential. Generally, rapid root elongation ceases well before the permanent wilting point. For example Newman (1966) found that growth of flax roots ceased at -0.7 MPa soil water potential in a sand-loam-peat soil and Taylor & Klepper (1974) found that cotton roots ceased growth at -0.1 MPa soil water potential in a fine sandy loam. Slow rates of root elongation may continue even in soils as dry as -4 MPa if part of the root system is in moist soil (Portas & Taylor, 1976).

The distribution of deep roots is influenced by the water content in the surface soil. For example, deep roots grow about 8 times as fast when surface soils are maintained at 20% of saturation than when they are maintained at 50% of saturation (Malik, Dhankar & Turner, 1979). Presumably available carbohydrate supplies are used up by proximal (surface) root growth and less is available for deeper root growth when the surface soil is wet enough to support rapid elongation.

These laboratory observations were confirmed in recent field experiments for wheat (R.K. Belford, unpublished data), sorghum (Blum & Ritchie, 1984), and three pasture grasses (Molyneux & Davies, 1983). The distribution of roots in field soils is often related to antecedent history of irrigation or to rainfall events. Fig. 7 shows the changed distribution of cotton roots caused by lack of irrigation (Klepper *et al.*, 1973). Generally well-watered crops in uniform soil profiles follow the rule-of-

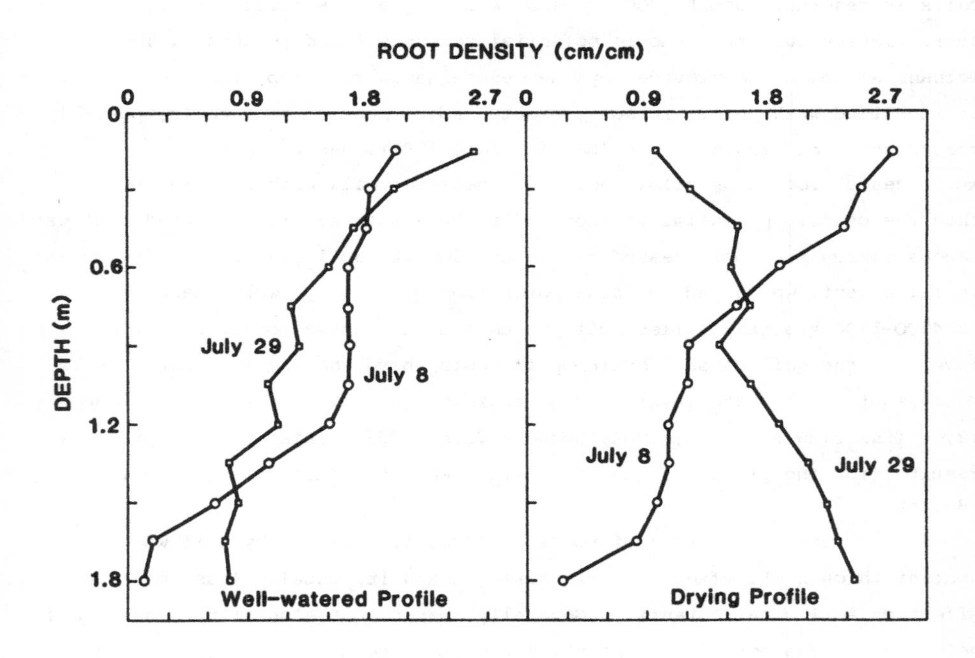

Fig. 7. Relationship between irrigation treatment and profile distribution of roots for cotton (*Gossypium hirsutum* L.). (Redrawn from Klepper *et al.*, 1973).

thumb that about 40% of the roots are in the top foot, 30% in the second foot, 20% in the third foot, and 10% in the fourth foot. Dryland crops have a more complex pattern with the distribution of roots being related to the weather of a particular year.

The distribution of roots at any time is determined by a balance between root growth and root death. The life span of individual root axes and branches has seldom been studied. It can range from a week or two for branch roots of apple and cotton (Huck, 1977) to several years for roots of trees and main axes of some perennial grasses (Troughton, 1981).

Factors other than soil water potential and soil strength influence root proliferation. Local concentrations of nutrient can have significant effects on roots. For example, the linear frequency of both lateral roots and main axes of maize increased by 50% and the rate of lateral root growth increased three-fold when nitrogen was increased from nil to 210 ppm (Maizlish, Fritton & Kendall, 1980). In the experiment, there was an 80% increase in the number of main axes and a 40% increase in their rate of elongation. In most experiments of this type, it is difficult to separate the immediate effects of nutrient-rich soil on the roots growing in that soil from the indirect effects of improved shoot nutrition on root growth. For example, the shoot dry weight in the experiment discussed above was heavier by a factor of nearly 20 in the 210 ppm treatment compared to the nil treatment.

Similar problems arise when studying effects of nutrient deficiencies on root growth. For example, in a study of two grasses, Christie (1975) found that a deficiency of phosphorus (0.003 ppm) resulted in a significant reduction in the length but not the number of lateral roots, and a great reduction in both the length and number of nodal roots when compared to plants grown with sufficient phosphorus (3.0 ppm); there was, however, a three-fold difference in shoot weight in these plants. Experiments show that isolated roots under laboratory conditions do indeed respond to local bands of enhanced fertility (Drew, 1975; Hackett, 1972).

Likewise, direct effects of root temperature on root elongation are difficult to separate from indirect effects on root growth associated with shoot response to root temperature. Table 2 shows the root dry weight responses obtained for eight crop species grown over a temperature range from 5°C to 40°C (Brouwer, 1962). Note that there is very little difference among the species in the upper temperature limit for root growth but considerable difference in the lower limit. Crops such as maize and beans have different lower limits from crops such as strawberry and peas. These cold effects are possibly related to species differences in effects of temperature on root permeability to water. Cool season crops show greater permeabilities to water at cold temperatures (Kramer, 1969). The same reductions in permeability which obtain during root function in water uptake for transport to the shoot possibly apply for the uptake of water into elongating tissues for growth.

Seasonal patterns of root distribution

For any field, temperature, water, air and nutrient distributions in the soil profile change during the growing season as a result of energy, gas and water exchange at the aerial interface of soil and plant surfaces, applications of fertilizers, plant uptake and soil release of nutrients, long-term drainage, and diffusion of heat, water, and nutrients within the soil profile. These complex dynamic shifts affect root growth and cause associated shifts in root distribution.

Each annual plant begins with a simple root system of very limited volume and eventually develops a large, complex network of roots by producing new meristems (branching) and maintaining old meristems (continued growth of main axes). The potential for vigorous growth on any day depends on the number of location of meristems generating cells for expansion on the previous day or two. As the season progresses, the growth of each root system is controlled by a complex interaction of prior crop history, shoot

Table 2. Root temperature range for growth. Data are from Brouwer (1962, Fig. 1) and have been obtained by interpolating to include the range of temperatures where growth is reduced to less than half of the maximum values.

Species	Temperature °C min	max
flax	10	31
peas	9	33
red kidney beans	12	33
maize	17	37
strawberries	5	31
broad beans	12	32
rape	16	32
oats	9	32

environment and physiology, and soil conditions.

Several authors have pointed out that the early growth pattern in a root system has a permanent effect on the future vigour and distribution of roots (Yorke & Sagar, 1970; May *et al.*, 1975). For example, if fertilizer applied at sowing burns one of the first three seminal roots of wheat, emergence is delayed because of the reduction in root area available for water uptake for coleoptile expansion (Klepper, Rasmussen & Rickman, 1983). Besides this delay in emergence, this loss of one seminal root is permanent. Neither the axis nor any of its branches will be available to extract water and nutrients from the soil as the seedling grows. Similarly, in some circumstances direct drilling of cereals may produce shallower seminal root systems than in ploughed soils and can result in a permanent reduction in the rooting depth of the seminal axes (Finney & Knight, 1973).

In crops with taproots (cotton or soyabean) loss or injury of the radicle will delay deep penetration of the taproot and its branches and will decrease the volume of soil explored. The function of the taproot will eventually be taken over by a strong lateral but only after significant delay. There is only one root system for many dicot seedlings and it arises from the radicle, making this root extremely important.

In contrast, cereals have several seminal roots and many individual crown roots associated with the main stem and its tillers. Injury of one of these is less drastic than it is in tap-rooted species. After the crown has been set, winter wheat begins to produce roots from the crown at about the same time that tillers begin to appear at the first foliar node (Klepper *et al.*, 1984). In wheat, crown roots grow at an oblique angle for some distance and then grow vertically after they are several centimetres from the centre of the plant (Weaver, Kramer & Reed, 1924). The initial angle or direction of growth appears to be related to diameter in maize (Yamazaki & Kaeriyama, 1982) and rice (Yamazaki, Morita & Kawata, 1981). When these crown roots encounter dry soil or other unfavourable conditions, they may die and force the plant to function using only the seminal root system, which is sufficient to carry a wheat plant to maturity (Passioura, 1972).

During vegetative development, the increase in root length or root dry weight is usually exponential over time. Individual root axes grow at rates which depend on a number of factors including root carbohydrate supply, temperature, and aeration status and the strength and water potential of the surrounding soil. Therefore, it is not surprising to find a range of values in the literature for root extension rates. For most crops under good conditions, roots grow at a rate of about 1 cm/day. For example, Ellis & Barnes (1980) report 0.5 cm per day for wheat roots during winter in England; Tennant (1976b) reports about 2 cm/day for the same species over most of the growing season in Australia. For maize plants, values of 11.0 cm/day have been reported (Kaeriyma & Yamazaki, 1983).

During reproductive development, some root growth continues if soil conditions permit. For example, Osmond & Raper (1982) found little detectable effect of reproduction on root development in tobacco because removal of the reproductive sink caused no response in root growth. Likewise Kaspar, Stanley & Taylor (1978) found that soyabean roots extended into deeper soil at rates well over 2 cm per day during early pod set and filling and that some new root material was observable until near physiological maturity. Böhm (1978) and Gregory *et al.* (1978) reported evidence of new root growth in wheat after anthesis. Root decomposition rates exceeded root growth rates in upper soil layers (Bohm, 1978) and new roots were produced below 80 cm (Gregory *et al.*, 1978) but nevertheless, in both studies, reproduction did not cause root growth to cease.

Viewed over the whole crop season, the root system accumulates dry matter vigorously relative to shoots early in the season and more slowly as the season progresses. For example, when rates of shoot and root dry matter accumulation of winter wheat were compared, the partitioning shifted in favour of shoots during the latter part of the season (Barraclough, 1984).

REFERENCES

Barraclough, P.B. (1984). The growth and activity of winter wheat roots in the field: root growth of high-yielding crops in relation to shoot growth. *Journal of Agricultural Science, Cambridge*, 103, 439-42.

Belford, R.K., Rickman, R.W., Klepper, B., & Allmaras, R.R. (1982). A new technique for sampling intact shoot-root systems for field-grown cereal plants. 1982 Agronomy Abstracts. p.11.

Blakely, L.M., Durham, M., Evans, T.A. & Blakely, R.M. (1982). Experimental studies on lateral root formation in radish seedling roots. I. General methods, developmental stages, and spontaneous formation of laterals. *Botanical Gazette*, 143, 341-52.

Blum, A. & Ritchie, J.T. (1984). Effect of soil surface water content on sorghum root distribution in the soil. *Field Crops Research*, 8, 169-76.

Böhm, W. (1978). Untersuchungen zur Wurzelentwicklung bei Winterweize. *Journal of Agronomy and Crop Science*, 147, 264-9.

Boone, F.R. & Veen, B.W. (1982). The influence of mechanical resistance and phosphate supply on morphology and function of maize roots. *Netherlands Journal of Agricultural Science*, 30, 179-92.

Brouwer, R. (1962). Influence of temperature of the root medium on the growth of seedlings of various crop plants. JAARB. I.B.S. 1962. pp. 11-18.

Brouwer, R. & Hoagland, A. (1964). Responses of bean plants to root temperatures. II. Anatomical aspects. JAARB. I.B.S. 1964. pp.23-41.

Burch, G.J., Smith, R.C.G. & Mason, W.K. (1978). Agronomic and physiological responses of soybean and sorghum crops to water deficits. II. Crop evaporation, soil water depletion and root distribution. *Australian Journal of Plant Physiology*, 5, 169-77.

Caradus, J.R. (1981). Root morphology of some white clovers from New Zealand hill country. *New Zealand Journal of Agricultural Research*, 24, 349-51.

Charlton, W.A. (1983). Patterns of distribution of lateral root primordia. *Annals of Botany*, 51, 417-27.

Cholick, F.A., Welsh, J.R. & Cole, C.V. (1977). Rooting patterns of semi-dwarf and tall winter wheat cultivars under dryland field conditions. *Crop Science*, 17, 637-9.

Christie, E.K. (1975). Physiological responses of semiarid grasses. II. The pattern of root growth in relation to external phosphorus concentration. *Australian Journal of Agricultural Research*, 26, 437-46.

Derera, N.F., Marshall, D.R. & Balaam, L.N. (1969). Genetic variability in root development in relation to drought tolerance in spring wheats. *Experimental Agriculture*, 5, 327-37.

Drew, M.C. (1975). Comparison of the effects of a localized supply of phosphate, nitrate, ammonium, and potassium on the growth of the seminal root system, and the shoot, in barley. *New Phytologist*, 75, 479-90.

Eavis, B.W. & Payne, D. (1969). Soil physical conditions and root growth. In *Root Growth*, ed. W.J. Whittington. pp. 315-38. London: Butterworth.

Edwards, C.A. & Lofty, J.R. (1978). The influence of arthropods and earthworms upon root growth of direct drilled cereals. *Journal of Applied Ecology*, 15, 789-95.

Ehlers, W., Köpke, U., Hesse, F. & Böhm, W. (1983). Penetration resistance and root growth of oats in tilled and untilled loess soil. *Soil and Tillage Research*, 3, 261-75.

Ellis, F.B. & Barnes, B.T. (1980). Growth and development of root systems of winter cereals grown after different tillage methods including direct drilling. *Plant and Soil*, 55, 283-95.

Finney, J.R. & Knight, B.A.G. (1973). The effect of soil physical conditions produced by various cultivation systems on the root development of winter wheat. *Journal of Agricultural Science, Cambridge*, 80, 435-42.

Friedman, R., Altman, A. & Zamski, E. (1979). Adventitious root formation in bean hypocotyl cuttings in relation to IAA translocation and hypocotyl anatomy. *Journal of Experimental Botany*, 30, 769-77.

Greacen, E.L. & Oh, J.S. (1972). Physics of root growth. *Nature, New Biology*, 235, 24-5.

Gregory, P.J. (1983). Response to temperature in a stand of pearl millet (*Pennisetum typhoides* S. & H.). *Journal of Experimental Botany*, 34, 744-56.

Gregory, P.J., McGowan, M., Biscoe, P.V., Hunter, B. (1978). Water relations of winter wheat. I. Growth of the root system. *Journal of Agricultural Science, Cambridge*, 91, 91-102.

Hackett, C. (1971). Relations between the dimensions of the barley root system: effects of mutilating the root axes. *Australian Journal of Biological Sciences*, 24, 1057-64.

Hackett, C. & Stewart, H.E. (1969). A method for determining the position controlled conditions, and some morphological effects of locally applied nitrate on the branching of wheat roots. *Australian Journal of Biological Sciences*, 25, 1169-80.

Hackett, C. & Bartlett, B.O. (1971). A study of the root system of barley. III. Branching pattern. *New Phytologist*, 70, 409-13.

Hackett, C. & Steart, H.E. (1969). A method for determining the position and size of lateral primordia in the axes of roots without sectioning. *Annals of Botany*, 33, 679-82.

Hoshikawa, K. (1969). Underground organs of the seedlings and the systematics of Gramineae. *Botanical Gazette*, 130, 192-203.

Huck, M.G. (1977). Root distribution and water uptake patterns. In *The Belowground Ecosystem: A Synthesis of Plant-Associated Processes*, ed. J.K. Marshall, pp. 215-26. Range Science Department Science Series No. 26, Colorado State University, Fort Collins.

Hurd, E.A. (1968). Growth of roots of seven varieties of spring wheat at high and low moisture levels. *Agronomy Journal*, 60, 201-5.

Kaeriyma, N. & Yamazaki, K. (1983). The development of rooting zone in soil in relation to the growth direction and the elongation rate of the primary roots in corn plants. *Japanese Journal of Crop Science*, 52, 508-14.

Kaspar, T.C., Stanley, C.D. & Taylor, H.M. (1978). Soybean root growth during the reproductive stages of development. *Agronomy Journal*, 70, 1105-7.

Klepper, B. (1983). Managing root systems for efficient water use: axial resistances to flow in root systems - anatomical considerations. In *Limitations to efficient water use in Crop Production*, ed. H.M. Taylor, W.R. Jordan & T.R. Sinclair, pp. 115-25. Madison, Wisconsin: American Society of Agronomy.

Klepper, B., Belford, R.K. & Rickman, R.W. (1984). Root and shoot development in winter wheat. *Agronomy Journal*, 76, 117-22.

Klepper, B., Rasmussen, P.E. & Rickman, R.W. (1983). Fertilizer placement for cereal root access. *Journal of Soil and Water Conservation*, 38, 250-2.

Klepper, B., Taylor, H.M., Huck, M.G. & Fiscus, E.L. (1973). Water relations and growth of cotton in drying soil. *Agronomy Journal*, 65, 307-10.

Köpke, U., Böhm, W. & Jachmann, T. (1982). Rooting patterns of three winter wheat cultivars in a field and greenhouse experiment. *Journal of Agronomy and Crop Science*, 151, 42-8.

Kramer, P.J. (1969). *Plant and Soil Water Relationships: A Modern Synthesis*. New York: McGraw-Hill.

MacKey, J. (1973). The wheat root. In *Proceedings 4th International Wheat Genetics Symposium*, Missouri Agricultural Experiment Station, Columbia, Missouri, pp. 827-42.

MacLeod, R.D. & Thompson, A. (1979). Development of lateral root primordia in *Vicia faba, Pisum sativum, Zea mays* and *Phaseolus vulgaris*: rates of primordium formation and cell doubling times. *Annals of Botany*, 44, 435-49.

Maizlish, N.A., Fritton, D.D. & Kendall, W.A. (1980). Root morphology and early development of maize at varying levels of nitrogen. *Agronomy Journal*, 72, 25-31.

Malik, R.S., Dhankar, J.S. & Turner, N.C. (1979). Influence of soil water deficits on root growth of cotton seedlings. *Plant and Soil*, 53, 109-15.
Mallory, T.E., Chiang, S.H., Cutter, E.G. & Gifford, E.M., Jr. (1970). Sequence and pattern of lateral root formation in five selected species. *American Journal of Botany*, 57, 800-9.
Martin, E.M. & Harris, W.M. (1976). Adventitious root development from the coleoptilar node in *Zea mays* L. *American Journal of Botany*, 63, 890-7.
May, L.H., Chapman, F.H. & Aspinall, D. (1965). Quantitative studies of root development. I. The influence of nutrient concentration. *Australian Journal of Biological Sciences*, 18, 25-35.
McIntosh, M.S. & Miller, D.A. (1981). Genetic and soil moisture effects on the branching-root trait in alfalfa. *Crop Science*, 21, 15-8.
Molyneux, D.E. & Davies, W.J. (1983). Rooting pattern and water relations of three pasture grasses growing in drying soil. *Oecologia*, 58, 220-4.
Newman, E.I. (1966). Relationship between root growth of flax (*Linum usitatissimum*) and soil water potential. *New Phytologist*, 65, 273-83.
Ney, D. & Pilet, P.E. (1981). Nutation of growing and georeacting roots. *Plant, Cell and Environment*, 4, 339-43.
O'Brien, L. (1979). Genetic variability of root growth in wheat (*Triticum aestivum* L.). *Australian Journal of Agricultural Research*, 30, 587-95.
Osmond, D.L. & Raper, C.D., Jr. (1982). Root development of field-grown flue-cured tobacco. *Agronomy Journal*, 74, 541-6.
Passioura, J.B. (1972). The effect of root geometry on the yield of wheat growing on stored water. *Australian Journal of Agricultural Research*, 23, 745-52.
Peterson, C.M., Klepper, B. & Rickman, R.W. (1982). Tiller development at the coleoptilar node in winter wheat. *Agronomy Journal*, 74, 781-4.
Portas, C.A.M., & Taylor, H.M. (1976). Growth and survival of young plant roots in dry soil. *Soil Science*, 121, 170-5.
Rickman, R.W., Klepper, B. & Belford, R.K. (1985). Developmental relationships among roots, leaves, and tillers in winter wheat. In *Wheat Growth Modelling*, ed. W.Day and R.K. Atkins, pp. 83-98. Plenum Publishing.
Riedacker, A., Dexheimer, J., Tavakol, R. & Alaoui, H. (1982). Modifications experimentales de la morphogenese et des geotropismes dans le systeme racinaire de jeunes chenes. *Canadian Journal of Botany*, 60, 765-78.
Riopel, J.L. (1969). Regulation of lateral root positions. *Botanical Gazette*, 130, 80-3.
Robertson, W.K., Hammond, L.C., Johnson, J.T. & Boote, K.J. (1980). Effects of plant-water stress on root distribution of corn, soybeans, and peanuts in sandy soil. *Agronomy Journal*, 72, 548-50.
Sasaki, O., Yamazaki, K. & Kawata, S. (1983). The relationship between the growth of crown roots and their branching habit in rice plants. *Japanese Journal of Crop Science*, 52, 1-6.
Sasaki, O., Yamazaki, K. & Kawata, S. (1984). The development of lateral root primordia in rice plants. *Japanese Journal of Crop Science*, 53, 169-75.
Shierlaw, J. & Alston, A.M. (1984). Effect of soil compaction on root growth and uptake of phosphorus. *Plant and Soil*, 77, 15-28.

Taylor, H.M. & Gardner, H.R. (1963). Penetration of cotton seedling taproots as influenced by bulk density, moisture content and strength of soil. *Soil Science*, 96, 153-6.

Taylor, H.M. & Klepper, B. (1974). Water relations of cotton. I. Root growth and water use as related to top growth and soil water content. *Agronomy Journal*, 66, 584-8.

Taylor, H.M. & Ratliff, L.F. (1969). Root growth pressures of cotton, peas, and peanuts. *Agronomy Journal*, 61, 398-402.

Tennant, D. (1976a). Root growth of wheat. I. Early patterns of multiplication and extension of wheat roots including effects of levels of nitrogen, phosphorus and potassium. *Australian Journal of Agricultural Research*, 27, 183-97.

Tennant, D. (1976b). Wheat root penetration and total available water on a range of soil types. *Australian Journal of Experimental Agriculture and Animal Husbandry*, 16, 570-7.

Troughton, A. (1981). Length of life of grass roots. *Grass and Forage Science*, 36, 117-20.

Troughton, A. & Whittington, W.J. (1969). The significance of genetic variation in root systems. In *Root Growth*, ed. W.J. Whittington, pp. 296-314. Proceedings of the Fifteenth Easter School in Agricultural Science, University of Nottingham, 1968. London: Butterworth.

Ueno, M. (1982). Root development and function of white clover in warmer region of Japan. *Japanese Agricultural Research Quarterly* 16, 198-201.

Vine, P.N., Lal, R. & Payne, D. (1981). The influence of sands and gravels on root growth of maize seedlings. *Soil Science*, 131, 124-9.

Weaver, J.E., Kramer, J. & Reed, M. (1924). Development of root and shoot of winter wheat under field environment. *Ecology*, 5, 26-50.

Whiteley, G.M. & Dexter, A.R. (1982). Root development and growth of oilseed, wheat and pea crops on tilled and non-tilled soil. *Soil and Tillage Research*, 2, 379-93.

Whiteley, G.M. & Dexter, A.R. (1984). Displacement of soil aggregates by elongating roots and emerging shoots of crop plants. *Plant and Soil*, 77, 131-40.

Yamazaki, K. & Harada, J. (1982). The root system formation and its possible bearings on grain yield in rice plants. *Japanese Agricultural Research Quarterly*, 15, 153-60.

Yamazaki, K. & Kaeriyama, N. (1982). The morphological characters and the growing directions of primary roots of corn plants. *Japanese Journal of Crop Science*, 51, 584-90.

Yamazaki, K. & Kaeriyama, N. (1983). The diameter of primary roots and the lateral root formation in corn plants. *Japanese Journal of Crop Science*, 52, 59-64.

Yamazaki, K., Morita, S. & Kawata, S. (1981). Correlations between the growth angles of crown roots and their diameters in rice plants. *Japanese Journal of Crop Science*, 50, 452-6.

Yamazaki, K., Sasaki, O. & Kawata, S. (1981). Correlations between the lateral root formation and the growth characteristics of crown roots in rice plants. *Japanese Journal of Crop Science*, 50, 464-70.

Yorke, J.S. & Sagar, G.R. (1970). Distribution of secondary root growth potential in the root system of *Pisum sativum*. *Canadian Journal of Botany*, 48, 699-704.

Yoshida, S., Bhattacharjee, D.P. & Cabuslay, G.S. (1982). Relationship between plant type and root growth in rice. *Soil Science and Plant Nutrition*, 28, 473-82.

GROWTH, RESPIRATION, EXUDATION AND SYMBIOTIC ASSOCIATIONS: THE FATE OF CARBON TRANSLOCATED TO THE ROOTS

H. Lambers

INTRODUCTION

In the past decade, we have gained considerable insight into the intricate details of photosynthesis in higher plants. Now the time has come to integrate this knowledge into the context of the whole plant. Which are the major sinks for the photoassimilates? And how are the strengths of these sinks affected by environmental conditions?

About half of all assimilated photosynthates is exported from the leaves to below-ground organs. Some of these photosynthates are respired there to generate metabolic energy for growth, maintenance and transport processes. A significant, but probably small, portion of the photosynthates may be exuded into the rhizosphere; these exudates include organic nitrogen compounds. In plants such as legumes, which live in symbiosis with nitrogen-fixing bacteria, some of the assimilates are utilized in growth and maintenance of *Rhizobium* and the reduction of atmospheric nitrogen. Other symbionts, such as mycorrhizae, may also consume a significant portion of photoassimilates.

The aim of this chapter is to evaluate the various carbon-requiring processes in the roots to estimate their relative size as a sink for energy and carbon.

EXUDATION

Many investigators have tried to quantify losses of soluble and insoluble organic compounds from the roots to the rhizosphere. The water-soluble compounds that are exuded by plant roots include reducing sugars, amino acids and amides (Bowen, 1969; Rovira, 1969; Ratnayake, Leonard & Menge, 1978; Graham, Leonard & Menge, 1981; Johnson *et al.*, 1982a,b). Table 1 shows the order of magnitude of the exuded compounds: 0.1-8.0 mg of reducing sugars and 0.04-1.2 mg of nitrogenous compounds per day and per g dry mass of roots. The large variation between the data is largely due to the physiological status of the plant, and only partly to the different species used by the various authors; at a low oxygen supply to the roots the release of exudates increases (Wiedenroth & Poskuta, 1981); also the phosphate concentration in the soil affects the rate of exudation (Bowen, 1969; Ratnayake *et al.*, 1978; Graham *et al.*, 1981). At low phosphate levels in the soil, the phospholipid concentration in the roots of *Sorghum vulgare* and *Citrus aurantium* is only half that of roots grown in soil with sufficient phosphate (Ratnayake *et al*,, 1978). Both the exudation of amino acids and the efflux of ^{86}Rb were correlated with the phospholipid concentration in the root tissue. This led Ratnayake *et al.* (1978) to the conclusion that the increased exudation under low phosphate conditions is due to the low level of phospholipids in the membranes of the root cells, which makes these membranes more permeable for many compounds, including reducing sugars and K^+. The significance of increased exudation at low phosphate concentrations is that micro-organisms such as mycorrhizae are encouraged to thrive in the rhizosphere resulting in an improved phosphate status of the plant (see below).

The phosphate concentration in the root is not the only factor

Table 1.

Exudation of reducing sugars and nitrogenous compounds (amino acids and amides) by roots of a number of plant species. Rates are "net losses", i.e. the difference between total loss and reabsorption by the plants. Values are expressed per gram dry weight of roots and per 24 h, even if originally expressed on a different basis

Species	μg reducing sugars	μg nitrogenous compounds	Special remarks	Reference
Citrus aurantium	100- 755	42- 120	Variation correlated with %P in roots	Ratnayake *et al.* (1978)
Citrus sinensis	277-7977	361-1236	Variation associated with light conditions	Johnson (*et al.* (1982a)
Sorghum vulgare	639-1296	92- 301	Variation correlated with %P in roots	Ratnayake *et al.* (1978)
Sorghum vulgare	2932-3804	126- 650	Four week old plants	Graham *et al.* (1981)

determining the rate of exudation. In *Chrysanthemum morifolium* the exudation of both reducing sugars and amino acids increased during flower-bud formation although the P content of the roots was constant. Increased exudation in these plants was ascribed to increased levels of the exuded compounds in the roots (Johnson *et al.*, 1982b).

With the soluble exudates, insoluble organic material is also lost from the roots. This may include mucilaginous compounds, sloughed-off cells and even parts of the root system that have died. André, Massimino & Daguenet (1978) found rates of root excretion of 3 mg (C) day^{-1} for 40-70 day maize plants; insoluble and soluble compounds contributed equally to the excretion. Expressed per g root dry weight, these rates range from 0.1 to 0.4 mg (C) day^{-1}. Other estimates (Snellgrove *et al.*, 1982) suggest that loss of insoluble matter may be an order of magnitude higher than that of soluble compounds. Barber & Martin (1976) attempted to quantify both soluble and insoluble losses from axenically grown (*i.e.*, without other organisms) wheat and barley plants. They expressed exudation as a portion of total photosynthesis and found considerably higher values than those mentioned above. However, because root respiration of their axenically grown plants must have been vastly underestimated (see Table 4 for more realistic estimates of root respiration), these estimates of the rate of exudation as a proportion of photosynthesis may be incorrect. Similar remarks pertain to an investigation by Whipps & Lynch (1983) using the same technique as Barber & Martin.

This leaves us only with the possibility of making a rough estimate of the amount of carbon atom exuded as a proportion of that fixed in photosynthesis. This estimation is based on information in Table 1 and the assumptions that the relative growth rate of the investigated plants was 10% per day, that for each carbon incorporated into dry-matter two were respired (see below), that the shoot:root ratio of the investigated plants was two, and that the amount of carbon lost in respiration in the tops was 5% of photosynthesis. It leads to an estimated loss of carbon as soluble exudates of 0.024-1.7%; the higher values referring to phosphate-deficient plants. Based on a labelling experiment with ^{14}C, Snellgrove *et al.* (1982) found values for 'root washings' of 0.2-0.6% of photosynthesis in *Allium porrum*. In the same experiment they found values for loss of insoluble carbon between 0.5 and 2.3% of photosynthesis in non-mycorrhizal plants. Thus, compared to utilization of carbon for growth and respiration (see below) the losses due to exudation are small.

Although the losses of carbon due to exudation may be small, they are likely to be of major importance for the functioning of a plant under adverse conditions, such as low available phosphate concentrations in the soil. The rate of exudation is correlated with the degree of mycorrhizal infection of the roots (Graham *et al.*, 1981; Johnson *et al.*, 1982b), probably due to increased growth of the fungus when more substrate leaks out of the roots. This increased growth might cause increased infection by the mycorrhizal fungus of the roots.

But also in roots of species which are not infected by mycorrhizae, such as many Proteaceae and some legumes, including *Lupinus albus* (Gardner, Parberry & Barber, 1981), exudation appears to be significant in improving the phosphate status of the plant. Such plants have "proteoid roots", bottlebrush-like clusters of rootlets, covered in a dense mat of root hairs (Purnel, 1960). The absorption zones of the individual roots overlap to a large extent so that these roots do not appear to be important for increased uptake due to increased surface area. Such structures are better suited to the retention of substances exuded by the roots (e.g. chelators) than to uptake of substances moving to the roots by diffusion. The latter is thought to be the major function of root hairs and mycorrhizae. Proteoid roots of *Lupinus albus* are also able to significantly lower the soil pH (from about 7 to 5), and to affect the reduction of manganese and some forms of iron (Gardner *et al.*, 1981; Gardner, Parbery & Barber, 1982a,b). Both activities, excretion of chelating compounds and lowering of the pH, may lead to the increased solubilization of phosphate. The exact nature of the exuded compound(s) in *Lupinus albus* is not known, but it involves a low molecular compound which associates with iron.

GROWTH

Carbon is translocated to the roots via the phloem, predominantly as sucrose. In the roots, part of the sucrose is transformed into cellular structures, e.g. protein, cellulose.

Following exposure of the youngest mature leaf of uniculum barley to $^{14}CO_2$, 93% of the label was exported within 24 h after exposure (Gordon, Ryle & Powell, 1977). Approximately 45% of all ^{14}C was lost in respiration 72 h after exposure. Most of the rest of the carbon was found equally in the roots and growing leaves. In *Allium porrum*, where the shoot to root ratio was about twice as high as that in uniculm barley, Snellgrove *et al.* (1982) found 57% of the label back in the shoot and 16% in the roots, 214 h after exposure of the entire shoot to $^{14}CO_2$. The rest of the label was lost in respiration. These estimates of the amount of photosynthate that is incorporated into root dry matter agree with those based on an investigation of the C-economy during several days of growth (Table 2): the values vary between 12 and 32%, although somewhat higher values have been found for plants that form a tap root (*Daucus carota*).

RESPIRATION

Respiration provides the driving force for biosynthetic reactions, for the maintenance of cellular structures and ion gradients, and for transport processes. We have a fairly good idea of the energy costs of biosynthetic reactions. Our understanding of the metabolic costs of maintenance processes is much scantier, and the current knowledge of the

Table 2.

Utilization of photosynthates for root growth as a percentage of the photosynthates produced daily. Estimates are based on experiments in which $^{14}CO_2$ was fed to a leaf or entire shoot (method A) or on an analysis of the C-economy of whole plants (method B).

Species	C-utilization for root growth	Method	Special remarks	Reference
Allium porrum	16	A		Snellgrove *et al.* (1982)
Daucus carota	17-35	B	Increases when taproot is formed	Steingroever (1981)
Hordeum vulgare	24	A		Gordon *et al.* (1977)
Lupinus albus	21	B	Non-nodulated plants; 55-65 days old	Pate *et al.* (1979)
Pisum sativum	12	B	"Equigrowing NO_3-fed"	De Visser (1984)
Plantago Lanceolata	20-28	B	Increases with	Lambers
Plantago major	22-32	B	age	*et al.* (1981)

costs involved in transport is only embryonic. In this section I will first provide information on the proportion of photosynthates that is lost in root respiration and subsequently try to estimate the relative sizes of the respiratory sinks in the roots.

In *Allium porrum* 18% of the $^{14}CO_2$ applied to the leaves was lost in root respiration, compared with only 6% in the leaves, despite them representing a greater proportion of dry weight. Other estimates are based on an analysis of the C-economy of plants (Table 3).

A combination of the information in Tables 2 and 3 provides the relative costs for root growth and root respiration. Additional information about these relative costs is included in Table 4; the cited papers do not allow expression of the carbon costs as a proportion of the plant's photosynthesis. To emphasize a point made above, data from Barber & Martin (1976) are also included: the relative size of the sinks for root growth and root

Table 3.
Utilization of photosynthates for root respiration as a percentage of the photosynthates produced daily.

Species	C-utilization for root respiration	Special remarks	Reference
Allium porrum	18	Non-mycorrhizal	Snellgrove *et al.* (1982)
Cucumis sativus	12-14		Challa (1976)
Daucus carota	14-19	Decreases with age	Steingroever (1981)
Lolium multiflorum	14-20	High at low light intensity	Hansen & Jensen (1979)
Lupinus albus	23	Non-nodulated; 55-65 days old	Pate *et al.* (1979)
Pisum sativum	18	"Equigrowing NO_3-fed"	De Visser (1984)
Plantago lanceolata	15-28	Decreases with	Lambers *et al.*
Plantago major	12-29	age	(1981)
Zea mays	18	56 days old	Massimino *et al.* (1980)

respiration in their paper is very different from all other data pertaining to a large number of species. Whipps & Lynch (1983), using the same technique, found similarly low values for respiration. These values are considerably lower than required to explain ATP-production for biosynthesis of root dry-matter (see later), so that Barber & Martin's (1976) technique, though potentially attractive, must be inappropriate.

Respiration for biosynthesis

Penning de Vries, Brunsting & Van Laar, (1974) made extensive calculations on the ATP-costs for the synthesis of the major compounds in plant dry matter. The calculations were based on data available from

Table 4.

C-utilization for root respiration as a proportion of the net amount of carbon translocated to the roots, i.e. the sum of carbon utilized for root growth plus that for root respiration. Only those references not included in Tables 2 and 3 are used. The *Senecio* plants had a low relative growth rate, due to mutual shading.

Species	C-utilization for root respiration	Special remarks	Reference
Helianthus annuus	0.67		Hatrick & Bowling (1973)
Helianthus annuus	0.35-0.44	Decreasing with increasing growth rate	Szaniawski & Kielkiewicz (1982)
Senecio aquaticus	>0.6	Low relative	Lambers &
Senecio jacobaea	0.5-0.7	growth rate	Steingroever (1978a)
Triticum aestivum	0.67	Young seedlings	Barneix, Breteler & Van de Geijn (1984)
Triticum aestivum	0.52	Sand-grown, nutrient-limited	Lambers *et al.* (1982a)
Zea mays	0.49		Veen (1980)
Hordeum vulgare	0.05-0.27	High values ascribed to microorganisms	Barber & Martin (1976)
Triticum aestivum	0.06-0.30		

biochemical handbooks. Combined with the biochemical composition of dry-matter, these values allow the calculation of the proportion of carbon lost in respiration for growth and carbon incorporated into dry-matter. The ATP-costs for synthesis of root dry-matter of the indicated composition is 21.54 mmol of ATP per g dry weight (Table 5, based on information provided by Lambers & Steingroever, 1978b). Assuming that 36 mol of ATP are produced in the oxidation of one mol of glucose, it is calculated that a minimum of 0.10 g of C is lost in respiration for every g of C incorporated into dry-matter. This is an absolute minimum for two reasons: (1) Carbon dioxide is not only lost in the processes of ATP-generation, but also in the production of C-skeletons; and (ii) the theoretical maximum efficiency of ATP production is assumed. Because additional respiration is required for the generation of "maintenance energy" (see next section) and for ion transport in the roots, it is obvious that the low values for respiration observed by Barber & Martin (1976) and Whipps & Lynch (1983) are about one order of magnitude

Table 5.

A simplified calculation of the ATP-costs for biosynthesis of the various compounds of root dry-matter. Calculations are based on data for roots of *Senecio aquaticus* (Lambers & Steingroever 1978b). ATP consumption is expressed in mmol per g compound or per amount of that compound present in root dry matter (last column). Transport costs for inorganic ions are not included in the calculations.

Compound	mg per g biomass	ATP consumption per g compound	ATP consumption
Inorganic ions	120	-	-
Organic acids	50	8.3	0.42
Soluble sugars	60	5.6	0.34
Inulin	10	18.6	0.18
Soluble polyuronic acids	35	17.8	0.62
Insoluble polyuronic acids	45	21.7	0.98
Hemicellulose	190	23.9	4.54
Cellulose	190	24.7	4.71
Lignin	80	27.8	2.22
Amino acids	60	0.0	0.00
Proteins	135	49.2	6.64
Lipids	25	35.4	0.89
Total	1000		21.54

too small.

Most experimental approaches to measure respiration for growth do not allow separation between respiration for growth and that for ion transport. This is one reason why experimental values are so much higher than that calculated above (Table 6). However, even where respiration for growth *strictu sensu* was determined, there was still a discripancy between experimental and theoretical values. This is largely due to the participation of the non-phosphorylating alternative path in root respiration.

Respiration for maintenance of root dry matter

Respiration is not only necessary to drive biosynthetic reactions which lead to an increment of dry matter, but also to maintain the cellular structures. The two major sinks for maintenance energy are thought to be protein turnover and the maintenance of gradients (*i.e.*, compartmentation of solutes).

Various methods have been used to separate the growth and maintenance components of respiration (for a review, see Lambers, 1985). A good approach seems to use plants in a non-exponential stage of growth

Table 6.

C loss in respiration for growth (including ion uptake). Values were calculated from literature data assuming 40% C in dry matter and a respiratory quotient of 1.0; they are expressed in mg C per g C incorporated into dry matter. Veen's value for *Zea mays* is treated separately, since his method separates respiration for growth *strictu sensu* and that for ion uptake; it is assumed that fresh maize roots contain 7% dry matter.

Species	C-loss	Reference
Helianthus annuus	450	Szaniawski & Kielkiewicz (1982)
Lolium multiflorum	350-610	Hansen & Jensen (1977)
Plantago lanceolata	1310	Lambers & Steingroever (1978b)
Senecio aquaticus	1340	Lambers & Steingroever
Senecio jacobaea	500	(1978a)
Zea mays	328	Veen (1980)

(Kimura, Yokoi & Hogetsu, 1978; Szaniawski & Kielkiewicz, 1982). Respiration of the plant, or preferably of the various organs separately, and the relative growth rate of the same plants or organs are measured during this growth stage. The values found for respiration are then plotted against those found for the relative growth rate and, in such a graph, the intercept with the y-axis gives the rate of maintenance respiration, and the slope of the line allows calculation of the respiration required for growth. Respiration for ion uptake will be part of the growth respiration in this approach (see next section for a more elaborate method, allowing further breakdown of components including one for ion uptake). Some values for the rate of maintenance respiration obtained with this method are included in Table 7. It should be borne in mind that the values cannot be transformed into moles of ATP consumed for maintenance, in the absence of information on the participation of different electron transport pathways in root respiration. However, for a full appreciation of the figures and to allow a comparison with estimated costs of protein turnover, it is assumed that only the cytochrome path contributes to respiration and that 36 mol of ATP are produced per mol of glucose.

No data are available on the rate of protein turnover in roots

Table 7.

Some values for the rate of maintenance respiration (in µmol of O_2 or CO_2 h^{-1} g^{-1} (dry weight) in roots, obtained with a method like that indicated in the text. Original data were transformed into data expressed on an ATP basis, assuming 36 mol of ATP produced and 6 mol of O_2 or CO_2 produced per mol of glucose oxidized.

Species	Maintenance respiration	Special remarks	ATP-requirement for maintenance	Reference
Helianthus annuus	25-104	Increases with temperature	150-624	Szaniawski & Kielkiewicz (1982)
Senecio aquaticus	260		1560	Lambers & Steingroever (1978a)
Senecio jacobaea	160		960	
Zea mays	14		84	Veen (1980)

and only very few data on protein turnover in leaves (Table 8). Taking an average value for protein half life of 110 h, a protein content in roots of 20% of dry matter, and the known costs for protein synthesis of 4 mol of ATP per mol of peptide bonds, the costs for protein turnover in roots are estimated as 0.04 mmol (ATP)h^{-1} g^{-1} (dry weight). This value is a factor of 2 (for maize) to about 15 (for other species) lower than those presented in Table 7. A small part of the difference can be explained by the participation of the alternative path in root respiration (see next section). It is impossible to estimate the costs involved in compartmenting solutes in the cell. Thus we can only account for a small fraction of the measured values of maintenance respiration.

It is also difficult to express the costs of maintenance respiration in the roots in terms of the plant's C-balance, because this changes so much with the growth rate and the shoot to root ratio of the plant. For the fast growing plants in Table 7, it amounts to approximately 15-50% of the amount of carbon utilized in respiration in the roots. Because 15-30% of all photosynthates are respired in the roots, a rough estimate of the maintenance costs of the roots is 10% of daily photosynthesis.

Respiration for transport

Three major transport events occur in plant roots: (i) uptake of nutrients from the surroundings, possibly followed by transport to the xylem, (ii) unloading of sucrose, and other compounds, from the phloem and subsequent delivery to the cells which use it for energy generation and the production

Table 8.
Half life of leaf proteins (in h) measured with the ^{3}H-acetic anhydride technique (Simpson, Cooke & Davies, 1981).

Species	Half life	Special remarks	Reference
Lemna spec.	73-143	Values decreased under N-deficiency	Davies (1979)
Lolium perenne	58-115	Values decreased with age	A.J. Barneix (unpublished)
Zea mays	130		Simpson *et al.* (1981)

of the appropriate C-skeletons, and (iii) unloading of compounds, such as amino acids and potassium, from the phloem followed by rapid loading into the xylem.

It would be presumptuous to suggest that we understand the costs involved in these transport processes. Only for nitrate uptake has one attempt been made to determine experimentally the costs for roots of *Zea mays* (Veen, 1980): they were estimated at 1.15 mmol (O_2) $mmol^{-1}$$(NO_3)$. This was approximately 38% of all respiratory costs in the roots and amounts to 19% of all carbon utilized in the roots. Considering the many uncertainties about fluxes and paths of ion transport, it is not yet possible to relate these costs to the stoichiometry of protons and ions other workers have found.

(i) Uptake of nutrients from the surroundings

Net uptake of ions by roots is determined by both the influx and efflux of these ions. We have recently come to realize that the efflux component can be quite large. In *Hordeum vulgare* the efflux rates for nitrate increased with increasing time of exposure to nitrate and could be of the same order of magnitude as the net uptake of this ion (Deane-Drummond & Glass, 1983). Also, in *Zea mays*, efflux (of phosphate) was found to be a significant component of net P uptake at concentrations comparable to those found in soil solutions (Elliot, Lynch & Lauchli, 1984).

The driving force for ion uptake is a proton gradient, so that the energy costs of ion uptake can primarily be expressed as the number of protons per ion absorbed. Ullrich-Eberius, Novacky & Van Bel (1984) suggested that phosphate uptake, energized by the electrochemical proton gradient, requires 2 H^+ per H_2PO_4 in *Lemna gibba*. If there is a quantitative analogy with mitochondria, this would be equivalent to approximately 2 molecules of ATP.

Knowledge of the stoichiometry of protons and ions is in itself not sufficient to calculate the costs of ion transport. It has been estimated that the maintenance of a membrane potential with a conductance of 1 S m^{-2} requires 27% of all turnover of ATP in the cell. Although the validity of such numbers has been questioned, they seem to be of the same order of magnitude as the values for animal cells (see Poole, 1978, and references therein for a thorough discussion of this subject).

More information on the costs for other ions, on the ATP-H^+-stoichiometry and on the influx-efflux ratio is required before such as those of Veen (1980) can be related to the more specific steps of ion uptake.

(ii) Unloading of sucrose from the phloem

Sucrose, translocated from the kernel via the phloem to the roots of *Zea mays* seedlings, moves via a symplastic pathway to the cells (Giaquinta *et al.*, 1983). Hydrolysis of sucrose, catalyzed by a soluble invertase, occurs mainly in the vacuole. A similar mechanism appears to operate in storage roots of *Beta vulgaris* (Wyse, 1979); in discs of these roots uptake of sucrose is against a concentration gradient and requires metabolic energy. Sucrose transport into protoplasts from developing soyabean cotyledons occurs via a proton co-transporting mechanism with a stoichiometry of approximately one proton per sucrose molecule (Lin, 1985). The loading of sucrose into the phloem is probably also via a sucrose/proton co-transport mechanism and the cost of sucrose transport across a leaf cell membrane has been calculated to be 1.1-1.4 molecules of ATP per molecule of sucrose (Giaquinta, 1983). If the mechanism of the unloading process in roots and its energetics are similar, 28-36 mg of C are consumed per g of C imported into the roots. However, these calculations are based on a number of assumptions needing further experimental verification.

(iii) "Cycling" through roots

Nitrogen, in the form of amino acids and amides, is transported in large quantities to the roots via the phloem. Some of this nitrogen is incorporated in root proteins, but significant quantities may be loaded in the xylem and translocated back to the shoot. In vegetative N-limited wheat plants, 79% of all N imported in the roots via the phloem was rapidly exported again (Simpson, Lambers & Dalling, 1982). In the grain-filling stage, a significantly greater proportion was translocated to the roots and immediately exported again (Simpson, Lambers & Dalling, 1983). Other ions, e.g. K^+ (A.J. Barneix, personal communication) are rapidly transferred from phloem to xylem in the roots. The degree of "cycling" through roots, at least of nitrogen, depends on the species under investigation and also largely on the environmental conditions, particularly the supply of water (Nicolas *et al.*, 1985) and nitrogen (Lambers *et al.*, 1982b).

The costs of cycling of amino acids and other ions are hard to quantify, in the absence of sufficient data on the extent of this process and of exact costs of transport. In wheat roots they are likely to be at least as great as the costs of net ion uptake under some investigated conditions (cf. Simpson *et al.*, 1982).

Respiration via the alternative pathway

Mitochondria of higher plants, more often than not, have a cyanide-resistant, alternative path, in addition to the cytochrome path (Day, Arron & Laties, 1980). The alternative path bypasses the proton extrusion sites beyond ubiquinone so that the ADP:O ratio of electron transport via this path is only 1 as compared to 3 if electrons are transferred from endogenous NADH to O_2 via the cytochrome path. In roots this alternative path is often, though not always, engaged and may be responsible for 50% of all oxygen uptake in roots.

The pathway generally acts as an "energy overflow", consuming an excess of carbohydrates translocated to the roots. However, De Visser (1984) presented results indicating that after addition of nitrate to an N-free nutrient solution, when the demand for ATP cannot be met by the cytochrome path, the alternative path may also play a role in the generation of ATP required for uptake of NO_3. (See Lambers, 1985, for a review of this subject.

If root respiration consumes 30% of all photosynthates, the alternative path, if responsible for half of the roots' respiration, consumes some 15% of all photosynthates. This is likely to be an upper limit. Some roots, e.g. those of N_2-fixing legumes (De Visser, 1984), hardly respire via the alternative path, so that a range of 0-15% of carbon consumed as a proportion of daily photosynthesis is indicated.

Environmental conditions, e.g. fluctuating salinity levels, N-supply, drought and varying light intensity, affect the engagement of the alternative path, generally in a manner as expected from the "energy overflow" hypothesis (Lambers, 1985).

SYMBIOTIC RELATIONS

Two different symbiotic associations will be discussed here. First, an association involved in fixation of atmospheric nitrogen, and second, mycorrhizae, which primarily function in the acquisition of phosphate, although other nutrients may be absorbed also.

N_2-fixing symbionts

Three different symbiotic associations involved in N_2-fixation are known in higher plants. The best documented is the *Rhizobium*-legume association, for which carbon costs are fairly well known. However, there are also associations between the actinomycete *Frankia* and a number of species, e.g. *Casuarina, Alnus, Myrica*. A third association is between a cyanobacteria (*Anabaena* spec. and *Nostoc*) and cycads, generally referred to as "coralloid roots". Bergersen, Kennedy & Wittman (1965) demonstrated fixation of atmospheric ^{15}N by coralloid roots of the Australian *Macrozamia communis*.

Carbon costs for the *Rhizobium*-legume symbiosis are for the synthesis and maintenance of nodules and for the reduction of atmospheric N_2 to NH_3. Pate, Layzell & Atkins (1979) found that 1056 mg C were lost as CO_2 from nodulated *Lupinus albus* plants (55-65 days old) compared with only 624 mg in NO_3-supplied, non-nodulated plants with a similar growth rate and of the same age. Of this 1056 mg, 354 mg was associated with nodule activity. For N_2-fixation *per se, i.e.* cost of nitrogenase activity and NH_3-assimilation, 299 mg C was used. The costs of NO_3-assimilation, which in *L. albus* predominantly occurs in the roots, in the NO_3-fed plants was 170 mg C per plant. N_2-fixation therefore was calculated to cost about twice as much as NO_3-assimilation. Most investigators find higher costs for N_2-fixation than for NO_3-assimilation. Based on data of Herridge & Pate (1977) and Pate *et al.* (1979), the costs of N_2-fixation of the *Rhizobium*-legume association are estimated as 4-12% of photosynthesis. The larger part, 80-90%, of the costs are the respiratory costs of the nodules associated with N_2-fixation *per se*; a small portion is associated with growth and maintenance of nodules.

Ryle, Powell & Gordon (1985a) found that in *Trifolium repens* plants, 23% of the photosynthates produced daily were required for N_2 fixation - a value considerably higher than those published by Pate and co-workers. Ryle, Powell & Gordon (1985b) ascribe the discrepancy to the great sensitivity of nodules to perturbation. This sensitivity is claimed to be responsible for a decrease of both N_2 fixation and respiration of the nodules during experimentation. The higher estimates of the carbon costs for N_2 fixation are therefore likely to be more realistic.

Mycorrhizae

Mycorrhizae are associations between soil fungi and plant roots which are often beneficial to plant growth, due to increased rates of nutrient absorption. The fungus depends on the plant for carbon supplies provided by photosynthesis, rather than by dead material from the soil (see Harley & Smith, 1983). Mycorrhizal associations are therefore a sink for assimilates and in recent years some investigations have been carried out to quantify the size of this sink.

Snellgrove *et al.* (1982) grew *Allium porrum* (leek) plants in partially-sterilized soil either inoculated or not with the vesicular-arbusenla mycorrhizal fungus, *Glomus mosseae*. By providing the non-mycorrhizal plants with more phosphate than the others, similar growth patterns were obtained for both groups of plants. Shoots were then pulse fed with $^{14}CO_2$ and, at the end of a chase period of 214 h, ^{14}C in shoot tissue, root tissue, soil organic matter, "root washings", and CO_2 from shoot and "below-ground" respiration was determined. Mycorrhizal plants assimilated the same amount of CO_2 but translocated about 7% more of the assimilates to below-ground parts. This was accounted for by increased root respiration (4.7%) plus increased loss of organic matter (presumably associated with mycelium formation; 2.8%). Less carbon was left in the shoot tissue, which had a lower percentage dry-matter than that of non-mycorrhizal plants.

Short term $^{14}CO_2$ labelling (8.5 min), followed by a 2 h chase, was carried out with *Citrus aurantium* (sour orange) and *Poncirus trifoliata* x *C. sinensis* (carrizo citrange)(Koch & Johnson, 1984). These plants were grown with a split root system, half of which was infected with *Glomus intraradices*, the other half being non-mycorrhizal. There was no significant difference in phosphate concentration between the two root halves. Two hours after labelling with ^{14}C, 6.3 and 3.3% of labelled photosynthates were present in mycorrhizal and non-mycorrhizal roots of sour orange. For citrange the proportions were 10.8 and 5.5%. The presence of the fungus, therefore, accounted for extra import of ^{14}C labelled assimilates.

The problem with such short chase periods, however, is that all the label, destined for the roots, will not yet have arrived by the sampling time. Moreover, some of the C may already have been lost in respiration. Ignoring this criticism, Koch & Johnson (1984) estimated the cost of sustaining mycorrhizae at 6% (sour orange) and 10.6% (carrizo citrange) of net photosynthesis in a plant with a root system which is totally mycorrhizal.

Although mycorrhizae tend to be beneficial for plant growth, they may also inhibit biomass production. This is probably due to competition for photosynthates. *Glycine max* plants were colonized by *Glomus fasciculatum* grown in media supplied with various levels of hydroxyapatite ($Ca_{10}(PO_4)_6(OH)_2$). At both high and low P supply the colonized plants showed growth retardation (20 and 38%, respectively). At intermediate P supply, growth was enhanced by 14% (Bethlenfalvay, Bayne & Pacovsky, 1983).

When *Phaseolus vulgaris* plants were grown symbiotically with both *Rhizobium phaseoli* and *Glomus fasciculatum* at levels of hydroxyapatite ranging from 0-200 mg per pot, dry mass was less than when grown with *Rhizobium* only (Bethlenfalvay *et al.*, 1982). Inhibition of growth and N_2-fixation was greatest at the highest and lowest P regimes and smallest when an intermediate level of 5 mg P was supplied per kg substrate.

It is concluded that growth inhibition was great in the absence of hydroxyapatite because *Glomus* could not have any beneficial effects but it could still consume carbohydrates. At high P supply, when mycorrhizae also do not have a beneficial effect on biomass production, and when the biomass of the endophyte was up to 10% of total (plant + symbionts) biomass, impaired growth and N_2-fixation was also thought to be due to competition for photosynthates between the symbionts.

CONCLUSION

Combining the information from the previous sections provides a summary of the relative sizes of carbon sinks in roots (Table 9). Growth and respiration are clearly the largest sinks for carbon in the roots. Symbiotic associations, at least under some conditions, are also major sinks, whereas exudation is probably less important. Taking into account that the

Table 9.
The fate of carbon translocated to the roots

Sink	Proportion of photosynthesis %	Degree of certainty "scale 0-10"
Exudation	5	4
Growth	11-35	9
Respiration*	12-29	8
- biosynthesis**	1- 4	7
- maintenance*	10	5
- ion uptake*	10	3
- other transport processes	3	1
- alternative path	0-15	5
N_2-fixation	5-23	5
Mycorrhizal symbiosis	7-10	5

*Values may include alternative path activity
**Based on theoretical calculations (Table 5).

alternative pathway plays a role in all respiratory processes included in Table 9, the five listed respiratory sinks seem to be of equal importance.

To indicate the reliability of these figures a "degree of certainty", ranging from 0-10, has been included. A low degree of certainty may indicate a paucity of data available from the literature, or a large disagreement between the available figures. Clearly more work needs to be done to appreciate the relative size of the respiratory sinks in the roots. Also the quantification of exudation warrants further research.

Table 9 does not give justice to the effects of environmental conditions on the relative size of carbon sinks in roots. Salinity, drought, nutrient stress, oxygen supply, light intensity and such atmospheric factors as the concentration of CO_2 and air pollutants all affect the pattern of carbon allocation. This may be via effects on growth, on respiration, on dry-matter partitioning between roots and shoots, on symbiotic associations or on exudation. A separate section on the carbon costs of living in adverse environments might have been included.

Genotypic differences, e.g. those associated with growth rate, have also been omitted from Table 9. Some knowledge is available on genotypic effects on carbon utilization in roots (cf. Lambers, 1985), but it is too early to provide an overall picture.

Techniques are now available to answer the sort of questions ecophysiologists put forward, such as what are the costs of living in a low nutrient environment? What could be the (dis)advantages of proteoid roots vs. mycorrhizal associations? etc. Such questions may be answered in the next few years.

ACKNOWLEDGEMENTS

Grassland Species Research Group, Publ. No. 103. I thank Hendrik Poorter for his constructive remarks on this manuscript.

REFERENCES

André, M., Massimino, D. & Daguenet, A. (1978). Daily patterns under the life cycle of a maize crop. II. Mineral nutrition, root respiration and root excretion. *Physiologia Plantarum*, 44, 197-204.

Barber, D.A. & Martin, J.K. (1976). The release of organic substances by cereal roots into soil. *New Phytologist*, 76, 69-80.

Barneix, A.J., Breteler, H. & Van de Geijn, S. (1984). Gas and ion exchanges in wheat roots after nitrogen supply. *Physiologia Plantarum*, 61, 357-62.

Bergersen, F.J., Kennedy, G.S. & Wittmann, W. (1965). Nitrogen fixation in the coralloid roots of *Macrozamia communis* L. Johnson. *Australian Journal of Biological Sciences*, 18, 1135-42.

Bethlenfalvay, G.J., Pacovsky, R.S., Bayne, H.G. & Stafford, A.E. (1982). Interactions between nitrogen fixation, mycorrhizal colonization, and host-plant growth in the *Phaseolus-Rhizobium-Glomus* symbiosis. *Plant Physiology*, 70, 446-50.

Bethlenfalvay, G.J., Bayne, H.G. & Pacovsky, R.S. (1983). Parasitic and mutalistic associations between a mycorrhizal fungus and soybean: The effect of phosphorus on host plant-endophyte interactions. *Physiologia Plantarum*, 57, 543-8.

Bowen, G.D. (1969). Nutrient status effects on loss of amides and amino acids from pine roots. *Plant and Soil*, 30, 139-42.

Challa, H. 1976. An analysis of the diurnal course of growth, carbon dioxide exchange and carbohydrate reserve content of cucumber. Netherlands: PhD Thesis, University of Wageningen.

Davies, D.D. (1979). Factors affecting protein turnover in plants. In *Nitrogen Assimilation of Plants*, ed. E.J. Hewitt & C.V. Cutting, pp.369-96. London: Academic Press.

Day, D.A., Arron, G.P. & Laties, G.G. (1980). Nature and control of respiratory pathways in plants; The interaction of cyanide-resistant respiration with the cyanide-sensitive pathway. In *The Biochemistry of Plants. A Comprehensive Treatise*, ed. P.K. Stumpf & E.E. Conn, pp. 197-241. London: Academic Press.

Deane-Drummond, C.E. & Glass, A.D.M. (1983). Short term studies of nitrate uptake into barley plants using ion-specific electrodes and $^{36}ClO_3^{-1}$ I. Control of net uptake by NO_3^{-1} efflux. *Plant Physiology*, 73, 100-4.

De Visser, R. 1984. Interactions between energy and nitrogen metabolism in Pisum sativum. Netherlands: PhD Thesis, University of Groningen.

Elliot, G.C., Lynch, J. & Lauchli, A. (1984). Influx and efflux of P in roots of intact maize plants. Double-labelling with ^{32}P and ^{33}P. *Plant Physiology*, 76, 336-41.

Gardner, W.K., Parbery, D.G. & Barber, D.A. (1981). Proteoid root morphology and function in *Lupinus albus*. *Plant and Soil*, 60, 143-7.

Gardner, W.K., Parbery, D.G. & Barber, D.A. (1982a). The acquisition of phosphorus by *Lupinus albus* L. I. Some characteristics of the soil/root interface. *Plant and Soil*, 68, 19-32.

Gardner, W.K., Parbery, D.G. & Barber, D.A. (1982b). The acquisition of phosphorus by *Lupinus albus* L. II. The effect of varying phosphorus supply and soil type on some characteristics of the soil/root interface. *Plant and Soil*, 68, 33-41.

Giaquinta, R.T. (1983). Phloem loading of sucrose. *Annual Review of Plant Physiology*, 34, 347-87.

Giaquinta, R.T., Lin, W., Sadler, N.L. & Franceschi, V.R. (1983). Pathway of phloem unloading of sucrose in corn roots. *Plant Physiology*, 72, 362-7.

Gordon, A.J., Ryle, G.J.A., & Powell, C.E. (1977). The strategy of carbon utilization in uniculm barley. I. The chemical fate of photosynthetically assimilated ^{14}C. *Journal of Experimental Botany*, 107, 1258-69.

Graham, J.H., Leonard, R.T. & Menge, J.A. (1981). Membrane-mediated decrease in root exudation responsible for phosphorus inhibition of vesicular-arbuscular mycorrhiza formation. *Plant Physiology*, 68, 548-52.

Hansen, G.K. & Jensen, C.R. (1977). Growth and maintenance respiration in whole plants, tops and roots of *Lolium multiflorum*. *Physiologia Plantarum*, 39, 155-64.

Harley, J.L. & Smith, S.E. (1983). *Mycorrhizal Symbiosis*. London: Academic Press.

Hatrick, A.A. & Bowling, D.J.F. (1973). A study of the relationship between root and shoot metabolism. *Journal of Experimental Botany*, 24, 607-13.

Herridge, D.F. & Pate, J.S. (1977). Utilization of net photosynthate for nitrogen fixation and protein production in an annual legume. *Plant Physiology*, 60, 759-64.

Johnson, C.R., Menge, J.A., Schwab, S. & Ting, I.P. (1982a). Interaction of photoperiod and vesicular-arbuscular mycorrhizae. *New Phytologist*, 90, 665-9.

Johnson, C.R., Graham, J.H., Leonard, R.T. & Menge, J.A. (1982b). Effect of flower bud development in *Chrysanthemum* and vesicular-arbuscular mycorrhiza formation. *New Phytologist*, 90, 671-5.

Kimura, M., Yokoi, Y. & Hogetsu, K. (1978). Quantitative relationships between growth and respiration. II. Evaluation of constructive and maintenance respiration in growing *Helianthus tuberosus* leaves. *Botanical Magazine (Tokyo)*, 91, 43-56.

Koch, K.E. & Johnson, C.R. (1984). Photosynthate partitioning in split-root citrus seedlings with mycorrhizal and nonmycorrhizal root systems. *Plant Physiology*, 75, 26-30.

Lambers, H. (1985). Respiration in intact plants and tissues: Its regulation and dependence on environmental factors, metabolism and invaded organisms. In *Encyclopedia of Plant Physiology*, Vol. 18, ed. R. Douce & D.A. Day, pp. 418-73. Berlin: Springer-Verlag.

Lambers, H. & Steingroever, E. (1978a). Efficiency of root respiration of a flood-tolerant and a flood-intolerant *Senecio* species as affected by low oxygen tension. *Physiologia Plantarum*, 42, 179-84.

Lambers, H. & Steingroever, E. (1978b). Growth respiration of a flood-tolerant and a flood-intolerant *Senecio* species: Correlation between calculated and experimental values. *Physiologia Plantarum*, 43, 219-24.

Lambers, H., Posthumus, F., Stulen, I., Lanting, L., Van de Dijk, S.J. & Hofstra, R. (1981). Energy metabolism of *Plantago major* ssp. major as dependent on the supply of mineral nutrients. *Physiologia Plantarum*, 51, 245-52.

Lambers, H., Simpson, R.J., Beilharz, V.C. & Dalling, M.J. (1982a). Translocation and utilization of carbon in wheat (*Triticum aestivum*). *Physiologia Plantarum*, 56, 18-22.

Lambers, H., Simpson, R.J., Beilharz, V.C. & Dalling, M.J. (1982b). Growth and translocation of C and N in wheat (*Triticum aestivum*) grown with a split root system. *Physiologia Plantarum*, 56, 421-29.

Lin, W. (1985). Energetics of sucrose transport into protoplasts from developing soybean cotyledons. *Plant Physiology*, 78, 41-45.

Massimino, D., André, M., Richaud, C., Massimino, J. & Vivoli, J. (1980). Evolution horaire au cours d'une journee normale de la photosynthese, de la transpiration, de la respiration foliaire et racinaire et de la nutrition N.P.K. chez *Zea mays*. *Physiologia Plantarum*, 48, 512-18.

Nicolas, M.E., Simpson, R.J., Lambers, H. & Dalling, M.J. (1985). Effect of drought on partitioning of nitrogen in two wheat varieties differing in drought tolerance. *Annals of Botany*, 55: 743-54.

Pate, J.S., Layzell, D.B. & Atkins, C.A. (1979). Economy of carbon and nitrogen in a nodulated and nonnodulated (NO_3-grown) legume. *Plant Physiology*, 64, 1083-8.

Penning de Vries, F.W.T., Brunsting, A.H.M. & Van Laar, H.H. (1974). Products, requirements and efficiency of biosynthetic processes: a quantitative approach. *Journal of Theoretical Biology*, 45, 339-77.

Poole, R.J. (1978). Energy coupling for membrane transport. *Annual Review of Plant Physiology*, 29, 437-60.

Purnel, H.M. (1960). Studies on the family Proteaceae. I. Anatomy and morphology of the roots of some Victorian species. *Australian Journal of Botany*, 8, 38-50.

Ratnayaka, M., Leonard, R.T. & Menge, J.A. (1978). Root exudation in relation to supply of phosphorus and its possible relevance to mycorrhizal formation. *New Phytologist*, 81, 543-52.

Rovira, A.D. (1969). Plant root exudates. *Botanical Review* 35, 35-57.

Ryle, G.J.A., Powell, C.E. & Gordon, A.J. (1985a). Defoliation in white clover: regrowth, photosynthesis and N_2 fixation. *Annals of Botany*, 56, 9-18.

Ryle, G.J.A., Powell, C.E. & Gordon, A.J. (1985b). Short-term changes in CO_2 evolution associated with nitrogenase activity in white clover in response to defoliation and photosynthesis. *Journal of Experimental Botany*, 36, 634-43.

Simpson, E., Cooke, R.J. & Davies, D.D. (1981). Measurement of protein degradation in leaves of *Zea mays* using (^{3}H)acetic anhydride and tritiated water. *Plant Physiology*, 67, 1214-9.

Simpson, R.J., Lambers, H. & Dalling, M.J. (1982). Translocation of nitrogen in a vegetative wheat plant (*Triticum aestivum*). *Physiologia Plantarum*, 56, 11-7.

Simpson, R.J., Lambers, H. & Dalling, M.J. (1983). Nitrogen redistribution during grain growth in wheat (*Triticum aestivum* L.) IV. Development of a quantitative model of the translocation of nitrogen to the grain. *Plant Physiology*, 71, 7-14.

Snellgrove, R.C., Splittstoesser, W.E., Stribley, D.P. & Tinker, B. (1982). The distribution of carbon and the demand of the fungal symbiont in leek plants with vesicular-arbuscular mycorrhizas. *New Phytologist*, 92, 75-87.

Steingroever, E. (1981). The relationship between cyanide-resistant root respiration and the storage of sugars in the taproot in *Daucus carota* L. *Journal of Experimental Botany*, 130, 911-9.

Szaniawski, R.K. & Kielkiewicz, M. (1982). Maintenance and growth respiration in shoots and roots of sunflower plants grown at different root temperatures. *Physiologia Plantarum*, 55, 500-4.

Ullrich-Eberius, C.I., Novacky, A. & Van Bel, A.J.E. (1984). Phosphate uptake in Lemna gibba Gl: energetics and kinetics. *Planta*, 161, 46-52.

Veen, B.W. (1980). Energy cost of ion transport. In *Genetic Engineering of Osmoregulation. Impact on Plant Productivity for Food, Chemicals, and Energy*, eds. D.W. Rains, R.C. Valentine & A. Hollaender, pp. 187-95. New York: Plenum Press.

Whipps, J.M. & Lynch, J.M. (1983). Substrate flow and utilization in the rhizosphere of cereals. *New Phytologist*, 95, 605-23.

Wiedenroth, E. & Poskuta, J. (1981). The influence of oxygen deficiency in roots on CO_2 assimilation rates of shoots and distribution of ^{14}C-photo-assimilates of wheat seedlings. *Zeitschrift für Pflanzenphysiologie*, 103, 459-67.

Wyse, R. (1979). Sucrose uptake by sugar beet tap root tissue. *Plant Physiology*, 64, 837-41.

DEVELOPMENT AND GROWTH OF ROOT SYSTEMS IN PLANT COMMUNITIES

P.J. Gregory

INTRODUCTION

Two features distinguish plants growing in communities from the isolated plants and parts of plants that are the subject of other papers at this symposium. Namely, they generally live in competition with each other and with other living organisms and, moreover, they most commonly have their roots in soil. The physical environment provided by mineral soils is determined primarily by the nature, composition and interaction of the solid, liquid and gaseous phases and is difficult to describe quantitatively. For example, changing the water content of a soil will result also in changes in its aeration, mechanical, thermal and nutrient-supplying properties, making relations between soil properties and root growth difficult to quantify. Several recent reviews and books have dealt with aspects of these relations (e.g. Arkin & Taylor, 1981; Feldman, 1984).

Clearly many factors affect root growth (for example, pH, Al toxicity, mineral nutrient deficiency) and although many of these will be individually of over-riding importance in specific circumstances, more generally, the principal soil physical factors affecting growth will be i) moisture status, ii) mechanical strength, iii) temperature, and iv) aeration status. Chemical and biological factors, such as the amounts and distribution of nutrients in the soil and whether the roots are infected with mycorrhiza will also have an important role in determining the development and growth of the root system, but these are largely outside the scope of this volume.

I shall examine the effects of the four principal physical constraints to root growth and briefly explore the interactions between the physical and biological environments and their effects on roots. Because it is often difficult to control physical variables in natural plant communities or crops, results are frequently extrapolated from experiments on a few plants grown in pots to the field; I shall indicate where this approach has been successful and suggest the direction of future research.

EFFECTS OF THE PHYSICAL ENVIRONMENT ON DEVELOPMENT AND GROWTH

Soil water

The water status of soils has a pronounced effect on the development, morphology and growth of roots. The effects on development are particularly evident in grasses which initiate root axes (nodal axes) towards the base of the stem. Fig. 1 shows results for an experiment with ryegrass grown in pots containing a sandy loam topsoil (Gregory, 1976). Keeping the top of the soil moist resulted in the rapid production of nodal axes and within 21 days all plants had nodal roots; where the topsoil was allowed to dry, however, only 25% of the plants had nodal roots. Although such extreme drying early in the life of the plant may be uncommon in many temperate climates,

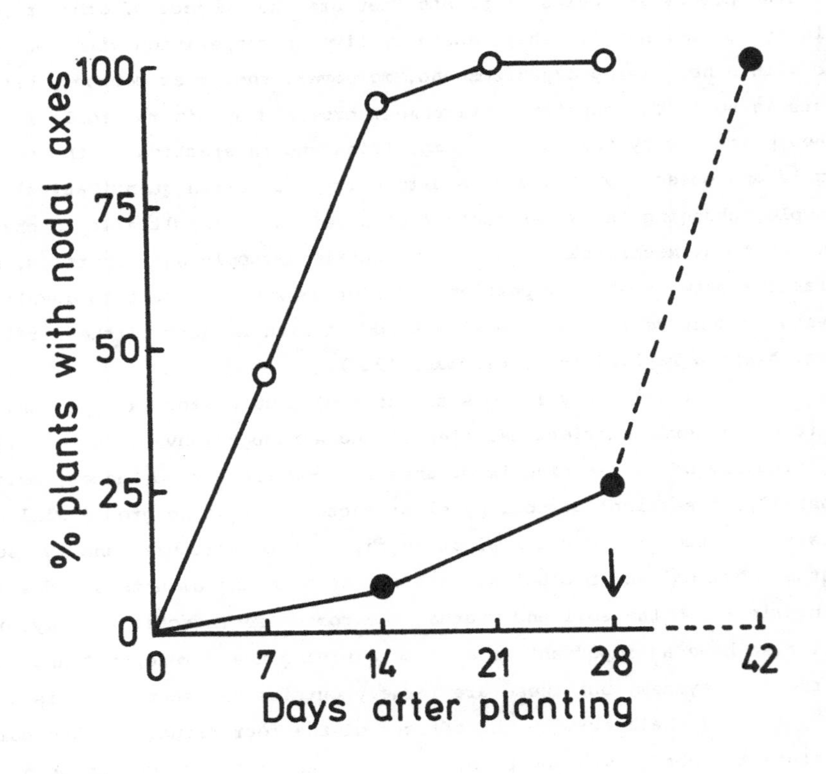

Figure 1. Effect of daily (O) and monthly (●) watering on the production of nodal axes by ryegrass. The arrow indicates the time of watering of the drier treatment.

Locke & Allen (1924) measured similar effects in a wheat crop grown in the USA and Cornish (1982) has indicated the detrimental effects that surface drying can have for the establishment of perennial grass seedlings in more arid environments. Re-watering of the plants resulted in the rapid appearance of new root axes (Fig. 1), a result also obtained by Troughton (1980) who investigated the response of clones of *Lolium perenne* (cv S.24) to drying of the soil surface. Troughton concluded that although the rate of development of new root axes after wetting varied genetically between varieties, the response was so rapid that it was doubtful whether selection for this character would improve the production of grasslands in temperate regions where prolonged periods of drying followed by wetting are commonly encountered in summer. However, in arid environments, the ability to produce new roots at the soil surface following rain may enhance drought resistance by making possible the transport of water to the plant that might otherwise be evaporated from the soil surface (Taylor & Klepper, 1978). Wilson, Hyder & Briske (1976) showed the importance of this attribute in the survival of blue grama grass (*Bouteloua gracilis*) grown in glasshouse and field experiments and Rogers, Unger & Kreitner (1984) have suggested a similar role for the adventitious roots of 'Hopi' sunflowers grown on sandy soils of the semi-arid and arid southwest USA.

Drying of the soil may also have marked effects on the visual appearance of roots. During drought, the production of mucigel is particularly noticeable, and substantial quantities of soil particles may adhere to the roots (Sprent, 1975). Mucigel production close to root tips may serve to lubricate the passage of roots through the soil and the mucigel bridge between root and soil particles during drought may contribute to continued water uptake. However, the precise significance of this last effect is not established.

In general, well-watered crops and plant communities have larger root systems (and shoots) than crops that are droughted and, furthermore, frequent light showers of rain, or frequent irrigation, encourage the proliferation of shallow roots. Where plants grow in soils that are wetted to considerable depth every year, many studies (e.g. Garwood, 1967; Cullen, Turner & Wilson, 1972; Merrill & Rawlins, 1979) show that the depth of rooting varies little in response to the frequency or amount of irrigation applied. However, in many soils, the depth of rooting is restricted mainly to the volume of soil that is regularly wetted so, for example, Weaver (1926) found that the rooting depth of winter wheat grown on a fine sandy loam/silt loam in the Great Plains varied from 150 cm to 120 cm to 60 cm when rainfall varied from

660-810 mm to 530-610 mm to 410-480 mm respectively. Even though the depth of rooting of annual plants may be unaffected by the frequency and depth of wetting, the length and distribution of roots most certainly are. Rowse (1974) found that the length of roots of mature field-grown lettuces was 75% less in plants protected from rainfall compared to plants that were frequently irrigated although root weights were similar. The non-irrigated plants had thicker roots and greater length and weight in the deeper soil layers. For winter wheat grown on a sandy loam soil (Gregory, 1976) irrigation affected both the total root length and the relative distribution of roots within the profile (Fig. 2).

In annual plants, the period of most active root growth occurs before the filling of the reproductive organs but, in perennial crops and pastures, a longer period of growth is often possible so that differences in growth, rooting depth and root distribution are more apparent between wet and dry years. Albertson (1937) showed such differences for prairie species

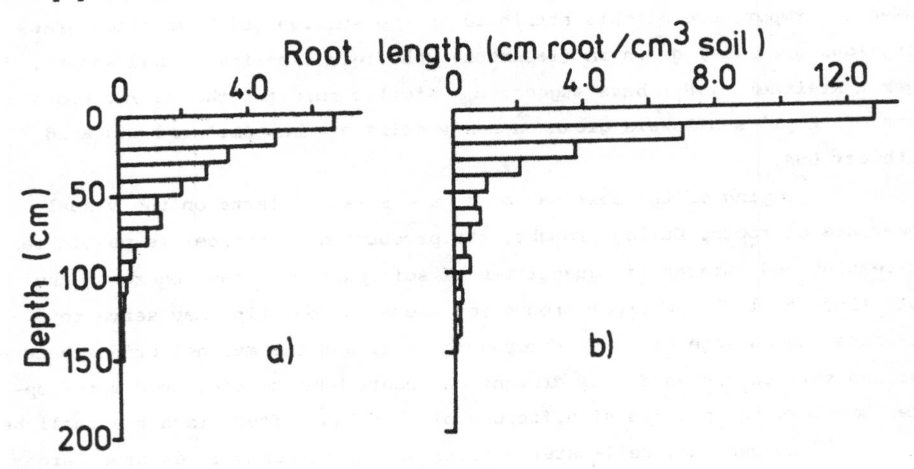

Figure 2. Distribution of root length at anthesis beneath winter wheat grown with a) normal rainfall, or b) irrigated to field capacity every week. The total dry weight of roots was a) 105 g/m^2, and b) 155 g/m^2.

and more recently Atkinson (1983) has summarized results showing considerable year-to-year variation in the production of white roots of apple trees.

Uncertainty exists as to how dry a soil must be before roots will not grow into it. Experiments in controlled conditions (Hendrickson & Veihmeyer, 1931; Portas & Taylor, 1976) suggest that roots can continue to grow in soils with matric potentials lower than the wilting point although at a very slow rate. In practice, soils with matric potentials as dry as -200 to -500 kPa are seen to restrict growth largely because soil matric potential affects plant turgor which in turn affects the ability of a root to overcome the mechanical resistance offered by the soil.

Mechanical resistance

The mechanical environment in which roots grow is extremely complex and differs not only between soils but also for the same soil at different water contents. The consequences of mechanical resistance or compaction are clearly visible in both natural ecosystems (see for example, Hosier & Eaton, 1980, where effects on dune and grassland vegetation were studied) and in agricultural systems especially where heavy machinery or inappropriate cultivation techniques are used (Taylor & Arkin, 1981). The roots of most plants have a diameter greater than 100 μm (see Barley, 1970, for diameter of cereal roots) and are therefore larger than the pores present in a compact, structureless soil. Field capacity in most soils of Southern England equates to a pore size of 60 μm (Webster & Beckett, 1972) so that a layer of soil can be well-drained yet have no pores large enough for roots to enter.

When a root growing in a structured soil comes into contact with the surface of an aggregate, three possible pathways may be followed. First, the root may penetrate the aggregate; second, the root may displace the aggregate from its path; or, finally, the root may bend to avoid the obstruction. The pathway actually followed will be a consequence of both the size and strength of the aggregate and the maximum buckling stress and growth pressure that the root can exert. Roots grow when the turgor pressure inside the elongating cells is sufficient to overcome the constraint imposed by the cell walls and any external constraint caused by the soil matrix. Thus, at equilibrium,

$$\Psi_i = \Psi_s + W + P$$

where Ψ_i is the water potential of the cells, Ψ_s is the water potential of the soil, W is the pressure exerted by the cell walls and P is the pressure (root growth pressure) exerted by the root against the soil. In some plants, Ψ_i can be modified by osmoregulation in response to changes of Ψ_s and mechanical resistance (Greacen & Oh, 1972). Gill & Bolt (1955) reviewed Pfeffer's work and found that the root growth pressure ranged from 700 to 2500 kPa for beans, maize, vetch and horse chestnut. Similarly, Taylor & Ratcliffe (1969a) showed that root growth pressures in cotton, peas and peanuts averaged 940, 1300 and 1150 kPa respectively although with considerable variation within each species. These pressures agree quite well with field-based assessments by penetrometry of the soil strength necessary to restrict root growth. Barley & Greacen (1967) reported that the range of mechanical resistance that stopped root growth was 800 to 5000 kPa depending upon plant species although Gerard, Sexton & Shaw (1982) reported root growth in coarse-textured soils with mechanical resistance of 6 to 7 MPa. Fig. 3 shows that the effects of increasing mechanical resistance on the elongation of cotton and peanut roots are qualitatively similar but quantitatively dissimilar. Penetrometers, though widely used, have a limited applicability in assessing the resistance offered to growing organs because not only is the pressure measured dependent upon the size of the penetrometer (Whiteley & Dexter, 1981) but rigid probes do not act physically like roots. Goss (1977) sought to overcome this problem by growing roots in a bed of glass beads that were subjected to an external pressure. He found that external pressures of only 20 and 50 kPa were sufficient to reduce the elongation of barley roots by 50% and 80% respectively. However, these external pressures should not be equated with the pressures experienced by the roots themselves because of arching or locking of the beads (Barley, 1963). Nevertheless, the experiments were useful in showing the effects of mechanical impedance on root morphology and anatomy. Increasing the external pressure to 20 kPa increased the diameter of the roots (mainly by increasing the thickness of the cortex), the number of cells in transverse section and the diameter of the outer cells, but decreased the diameter of the inner cells and the rate of root elongation resulting in shorter, thicker, roots (Wilson, Robards & Goss, 1977; see, however, chapters 1 and 2 for comment on these observations). More recently, Whiteley & Dexter (1984a) have made a theoretical analysis of the displacement of aggregates by elongating roots in relation to the degree of aggregate displacement (Δ) and root deflection (δ). When a root meets an aggregate, the amount of root deflection will equal the amount by which the aggregate is not deflected (i.e. $\Delta + \delta = c$, where c is a constant). For example, with $c = 0.3\ D$ (where D is the diameter of the aggregate), Whiteley & Dexter calculated that significant displacement of aggregates at 20-100 mm depth can only occur where the diameter of the aggregate is less than 1 mm and the diameter of the roots is less than 0.5 mm. Although such calculations reflect only approximately the true conditions in the topsoil, they are nevertheless useful indications of the likely behaviour of crop roots in seedbeds and of the importance of the diameter of roots when assessing the competitive ability of plants to establish themselves in natural communities.

In many soils, roots are confined to the planes of weakness between the soil peds (Taylor & Arkin, 1981) so that soil-root contact

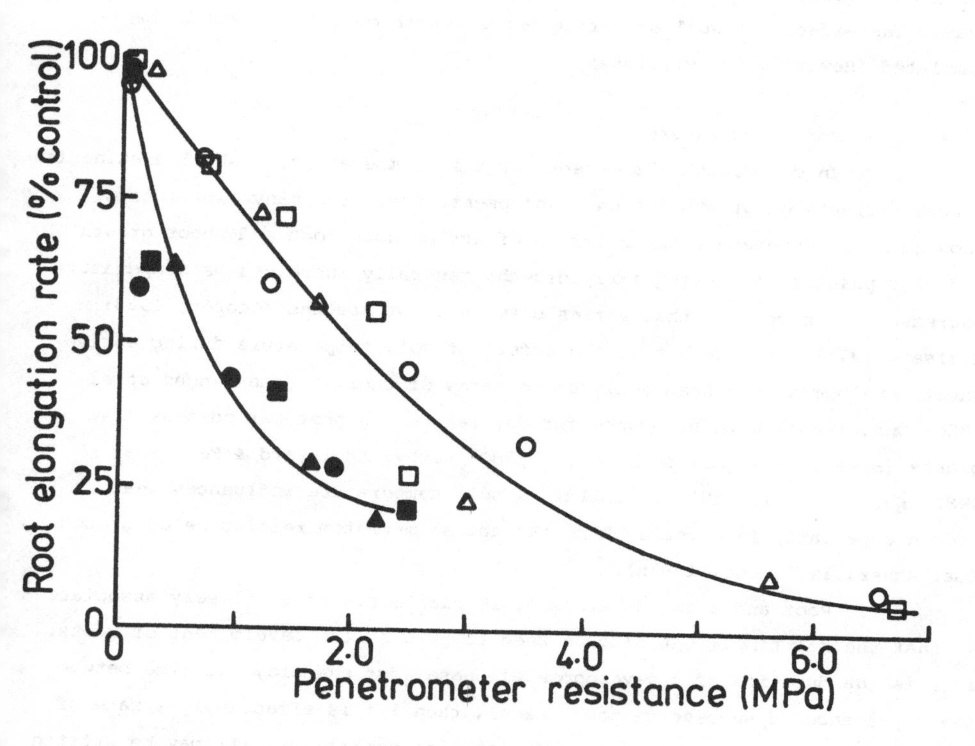

Figure 3. Effect of penetrometer resistance on the rate of root elongation of peanut (open symbols) and cotton (closed symbols) for the period 40 to 80 hours after transplanting, approximate water contents (g/g) of soil were 7% (circles) 5% (triangles) and 4% (squares). Redrawn from Taylor & Ratcliff (1969b).

and exploitation of slowly diffusible ions may be limited to only a small fraction of the total soil volume. Cracks and large pores allow roots to grow through soil layers that would otherwise provide great mechanical resistance so that the effects of such layers are often transient. For example, manually created pans in two clay soils were readily lost on drying through the formation of cracks; the pans had no effect on yields of ryegrass, spring barley or field beans despite transient effects on water uptake by roots (McGowan, Wellings & Fry, 1983). Whiteley & Dexter (1983 and 1984b) have investigated the growth of pea, rape and safflower roots in cracks under controlled conditions. The strength of the peds, the width of the crack, and the orientation of the crack relative to the preferred geotropic growth direction of the roots were all found to affect growth. Probabilities of root growth bridging cracks have been estimated and penetration criteria based on effective penetrometer/root stress ratios have been developed to enable the effects of soil structure and strength on root growth to be simulated (Hewitt & Dexter, 1984).

Soil Temperature

In describing the effects of soil temperature, I shall distinguish between effects on growth and on development. There are many experiments showing that temperature has a marked effect on both root and shoot growth and that plant size, during early growth, generally increases as temperature increases to an optimum that varies between plant species (Cooper, 1973; Nielsen, 1974). More recently, the effect of soil temperature during early shoot development has been analysed in terms of thermal time (Angus *et al.*, 1980; Ong, 1983a) with allowance for day length in photoperiod-sensitive plants (Baker, Gallagher & Monteith, 1980; Kirby, Appleyard & Fellowes, 1982; Hadley *et al.*, 1984). Similarly, soil temperature influences leaf growth especially in cereals where the apical meristem remains below ground (Gallagher, 1979; Ong, 1983b).

Root and shoot development in plants are often closely associated so that thermal time might also be used to predict the development of roots. If t is the duration of a developmental phase (for example, the time between the appearance of successive nodal axes), then $1/t$ is effectively a rate of development which, if it is linear function of temperature (T) may be written

$$1/t = (T - Tb)/\theta$$

where T_b is the base temperature at which the rate is zero and θ is a constant. Measurements by Nishiyama (1976) on rice seed radicles can be plotted as a linear relation between $^1/_t$ for the initial elongation to 5 mm (Fig. 4); the author originally published these data as an Arrhenius plot and suggested changes in activation energy at 17.5°C. Root systems of plants growing in soils are normally subjected to a range of temperatures

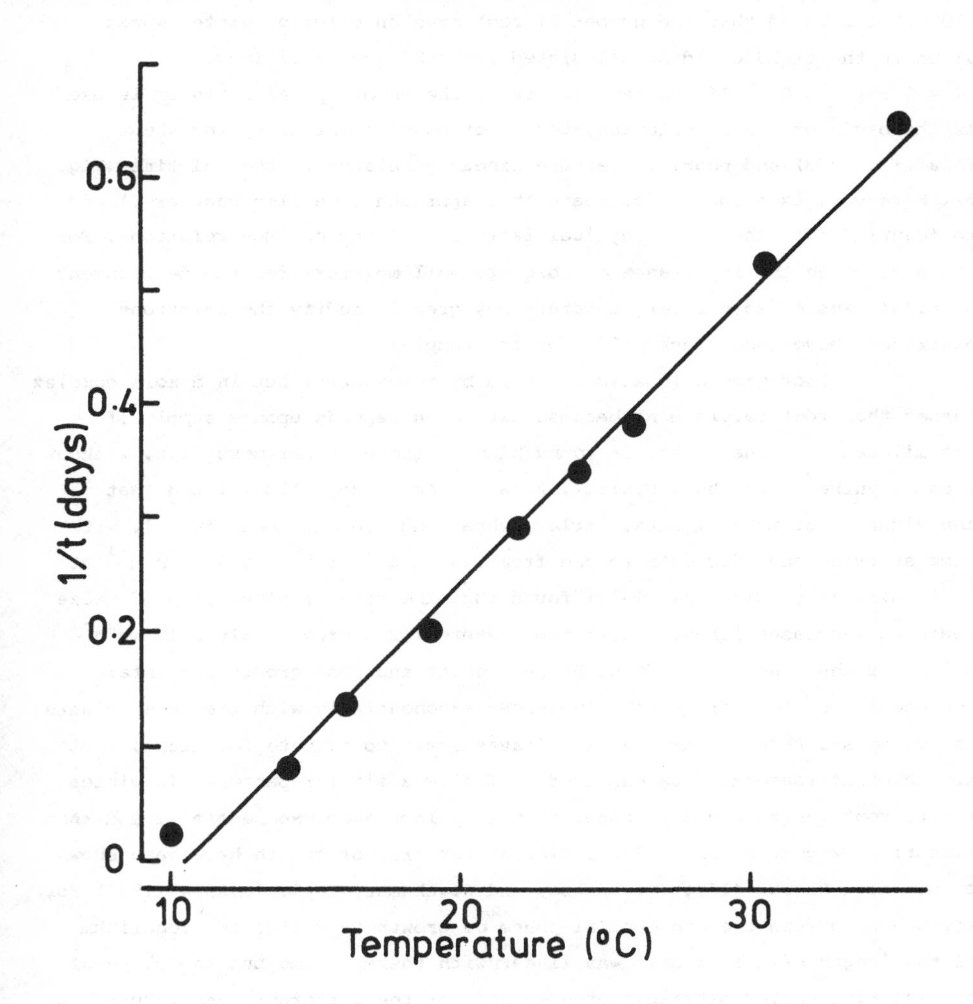

Figure 4. Effect of temperature on the rate of development of rice roots. The time (t) is the number of days for 50% of the seed roots to elongate to 5 mm, and the line drawn is the linear regression (y = 0.025 x - 0.266; r^2 = 0.99). After Nishiyama (1976).

so that the effective temperature, T is often difficult to identify. For example, Gregory (1983) found such difficulty in determining T_b for the production of root axes in millet plants that he resorted to using leaf number as a measure of thermal time. The number of axes (y) per plant was linearly related to the number of leaves per plant (x) for fourteen different temperature treatments ($y = 1.42x - 2.26$). Klepper, Belford & Rickman (1984) also found that the number of root axes on culms of winter wheat grown in the field could be calculated from the number of leaves ($y = 1.95x - 3.06$). If the temperature of the shoot apical meristem is used as the basis of the calculation, then root development of spring wheat (Alafifi, 1983) and pearl millet are linearly related to thermal time (Fig. 5). More work is needed to validate this approach (see also chapter 5) and to identify the other soil physical factors influencing such relations. For example, given the importance of adequate soil moisture for the development of nodal axes cited earlier, moisture may greatly modify the relations described above (see Black, 1970 for an example).

Root growth is also affected by temperature but in a more complex manner than root development because extension depends upon a supply of assimilates from the shoot. In young plants, grown in darkness (i.e. without a photosynthesizing shoot system), Abbas Al-Ani & Hay (1983) found that individual root axes of oats, barley, wheat and rye extended linearly with time at rates that for oats ranged from 0.4 cm d^{-1} at 5°C to 7.5 cm d^{-1} at 25°C. Similarly, Blacklow (1972) found that the rate of elongation of maize radicles increased linearly with temperature (at approximately 0.16 mm h^{-1} K^{-1}) over the range 10 to 30°C. Because shoot and root growth are interdependent and shoot dry weight increases exponentially with time when plants are young and then linearly as the leaves start to compete for light, root growth might reasonably be expected to follow a similar pattern. In winter wheat, root length and dry weight initially increased exponentially and then linearly (Gregory *et al.*, 1978). Similar patterns of growth have been shown by Mengel & Barber (1974) for maize, and Sivakumar, Taylor & Shaw (1977) for soyabeans. During the exponential phase of growth of millet the logarithm of the length of a root axis was linear with thermal time but the slope of the relation varied between root axes and for the different temperature treatments applied (Gregory, 1986). This marked effect of ontogeny on the elongation rate of individual root axes is similar to that found between successive leaves (Gallagher, 1979). However, Stone & Taylor (1983) found only very small differences in the rates of extension of taproots and lateral

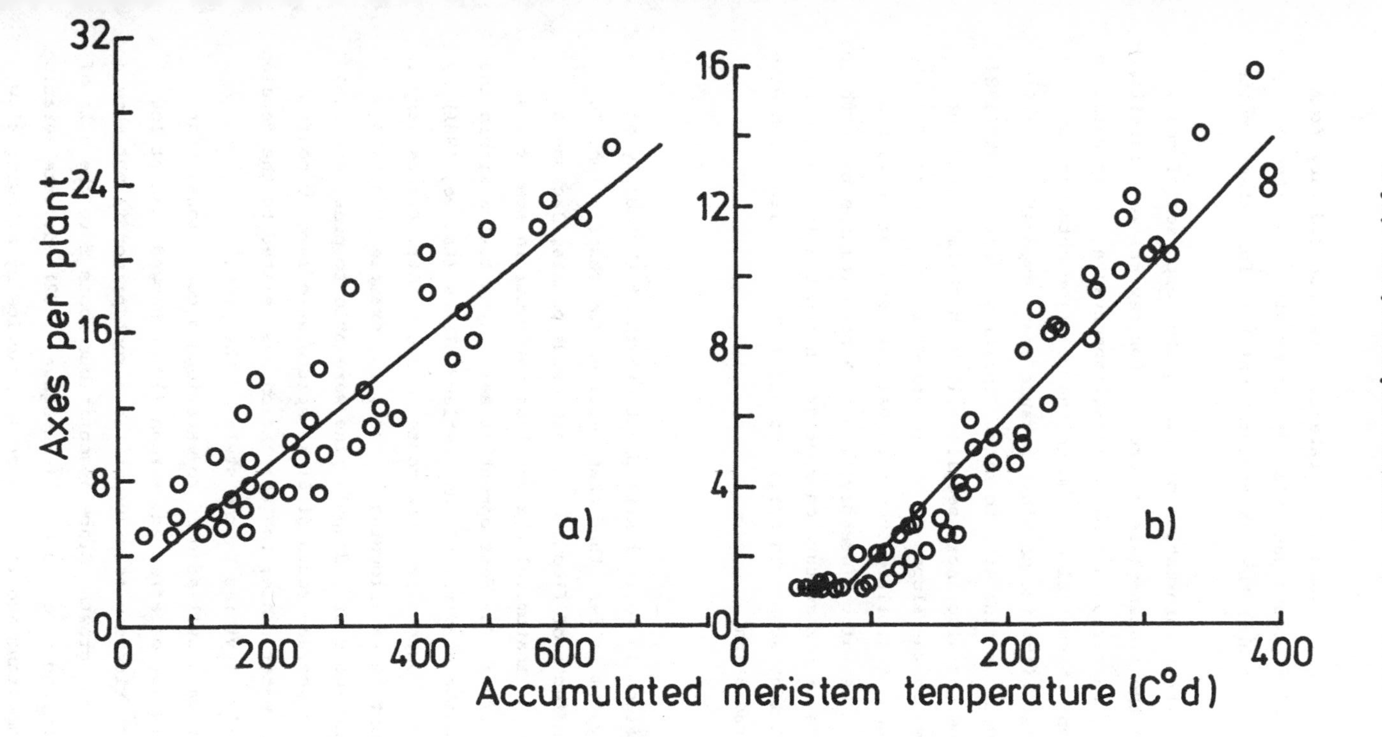

Figure 5. Relations between the number of root axes and thermal time measured at the shoot meristem for a) spring wheat (base temperature 0°C) and b) pearl millet (base temperature 12°C). The points represent soil temperatures ranging from 6-24°C for a) and 19-31°C for b). Redrawn from Alafifi (1983) and Gregory (1983).

roots in soyabean. Moreover, although the rate of extension did vary from day to day, the rate for taproots generally decreased with time (the decrease being greater the higher the temperature) but that for lateral roots remained constant.

There is little information describing the responses of roots to temperature during the linear phase of growth. For several well-fertilized crops of winter wheat, Barraclough & Leigh (1984) observed a linear relation between root length and thermal time although the relation obtained was little better than that with time possibly because mean temperatures varied little from year to year or from site to site. Similarly, Barraclough (1984) found that root dry weight also increased linearly with thermal time but there were differences in the slopes of the relations between years. Clearly, root growth depends upon assimilate supply as well as temperature so that these relations cannot be unique (Gregory, 1986). A need remains to establish more fully the effects of temperature on root growth and the interactions with the rates of nutrient supply in soils (Schaff & Skogley, 1982) and with the interception of radiation.

Aeration

Poor aeration in soils limits plant growth if the supply of oxygen to the root system is less than that required for respiration. In their review on the effects of flooding on herbaceous plants, Jackson & Drew (1984) describe the anatomical changes that may occur in some roots that alleviate such stress. The development of aerenchyma tissue within the cortex contributes to the survival of many wetland plants (Kawase, 1981) by allowing the transfer of oxygen from the roots to the shoots. Plants vary in their capacity to adapt to waterlogging so that, for example, Coutts & Philipson (1978) found roots of Lodgepole pine were able to penetrate into soils devoid of oxygen whereas roots of Sitka spruce were largely confined to the soil above the water table; the adaptation was related to the development of large gas-filled cavities in the stele of the pine.

Even in plants well-adapted to waterlogged conditions, root development and growth may be affected. Kordan (1976) showed that at low oxygen concentrations, rice may form adventitious root primordia but show no visible adventitious root growth. Since adventitious roots form the bulk of the root system in lowland rice (Yoshida & Hasegawa, 1982) adequate aeration of paddy fields is important and is achieved in practice by allowing slow percolation of water.

In arable crops, the effects of transient waterlogging on yields have been examined by Belford *et al*. (1980) for peas, Cannell *et al*. (1980) for winter wheat and Cannell & Belford (1980) for oilseed rape. The results show that the timing and duration of waterlogging in relation to crop growth stages, and the height of the water table all affected the final response which varied from nil for early winter and late spring waterlogging of rape, and for mid-winter waterlogging of wheat, to a 42% reduction in the yield of peas when waterlogged just before flowering.

ROOTS AND COMPETITION BETWEEN PLANTS

Differences between and within species

The morphology and growth rates of root systems are known to differ considerably between plant species (see for example, Kutschera, 1960) and these differences may relate to the successful exploitation of particular environments. Using Grime's (1979) ideas of stress-tolerant, competitive and ruderal species, Chapin (1980) linked together aspects of root growth, absorption capacity for nutrients and soil fertility to assess the strategies likely to be successful in various natural plant communities. Numerous characteristics of root systems also differ within species. In wheat, differences in the number of seminal roots (Robertson, Waines & Gill, 1979) the diameter of xylem vessels (Richards & Passioura, 1981) and the pattern of root growth (Hurd, 1974; O'Brien, 1979) have been observed and similar variation exists within other crop species (Caradus, 1976, clover; McIntosh & Miller, 1980, alfalfa; Armenta-Soto *et al*., 1983, rice). However, there have been few quantitative studies showing that these differences in rooting are important either for competition between plant species or for allowing one genotype to grow better than another at a specific location. Fig. 6 shows the root length between anthesis and maturity of two varieties of barley grown on a red vertisol (Palexerollic Chromoxerert-U.S.D.A.) in northern Syria (based on Brown, 1984). Although the two varieties had very similar total root lengths, the distribution of root length differed with Arabic Abiad having shorter length in the surface layer (0-15 cm) but longer length in layers below 30 cm. This greater length at depth was associated with more water extracted from these layers and a higher final dry matter and grain yield. Nevertheless, despite these correlative studies, specifying the shape and size of root systems for specific functions remains for the future. Models of root growth and function (Fowkes & Landsberg, 1981; van Noordwijk, 1983) may be useful in stimulating further developments in this subject.

Biological aspects

In addition to the gross differences between root systems, there are also differences between species brought about by the various symbiotic and pathogenic relations between roots and other biological organisms. Several recent publications (e.g. Sanders, Mosse & Tinker, 1975; Atkinson *et al.* 1983; Tinker, 1984) have described mycorrhizal symbioses with roots and their importance especially for phosphate and trace element (particularly Zn) uptake. By extending into the soil around the root, the mycorrhizal hyphae effectively shorten the distance over which slowly diffusible ions, such as phosphate, must travel before being taken up. The external hyphae absorb phosphate from the same pool of nutrients as the roots (the labile pool), and bypass the restriction to diffusion in the same manner as suggested for root hairs. However, the association is not without an energy cost to the host. Pang & Paul (1980) and Snellgrove *et al.* (1982) suggest that 4-10% of host photosynthate translocated below ground may be utilized in mycelium growth and respiration. Several reports of reduced yields after infection with vesicular-arbuscular mycorrhiza are possibly the result of this energy

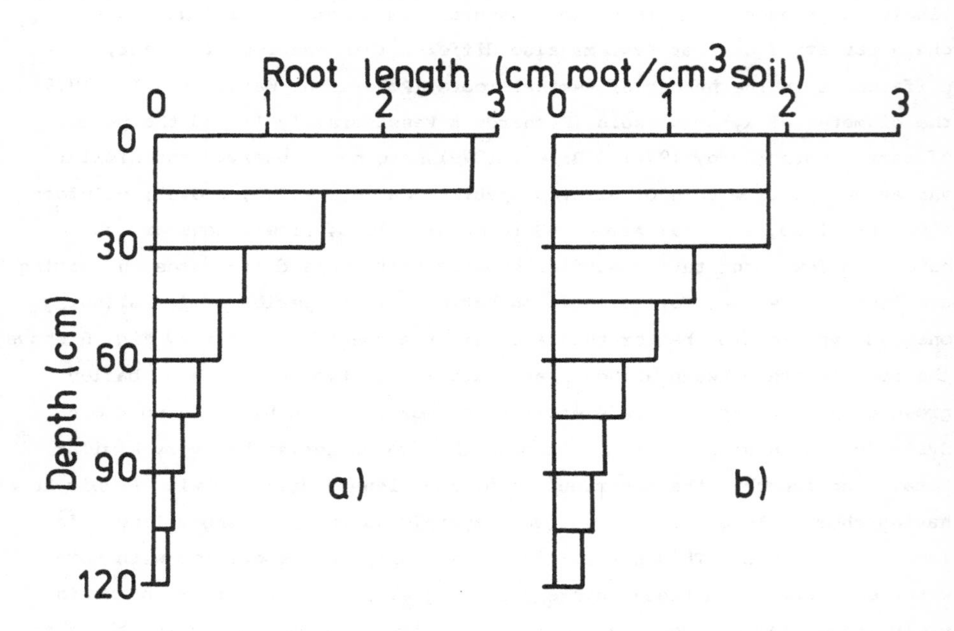

Figure 6. Distribution of root length between anthesis and maturity for 2 varieties of barley (a) Beecher and b) Arabic Abiad) grown at Jindiress in N.W. Syria. After Brown (1984).

cost (Buwalda & Goh, 1982).

The development of mycorrhizal associations is particularly beneficial to the host in soils where the concentrations of available P are low; increasing rates of phosphorus fertilizer decrease mycorrhizal infection (Sanders & Sheikh, 1983). The widespread, practical exploitation of the symbiosis is most likely to occur in low-input agriculture where the cost of phosphate fertilizer is a major limitation. However, our knowledge is as yet insufficient for this to be a practical possibility (Abbott & Robson, 1982). The ecology of the fungi is poorly understood and the effects of the physical environment have hardly been investigated. Infection does not occur in waterlogged soils and is also impaired by drought; Reid & Bowen (1979) found that infection at a matric potential of -190 kPa was almost twice that at saturation and four times that at -430 kPa.

Roots also may affect the ability of plants to compete with each other through the production of allelopathic chemicals (Rice, 1984). The precise modes of action of these chemicals is still being researched but their production seems to offer explanations for the commonly observed patterns of species in natural ecosystems and for certain herbicidal properties in agriculture. For example, sorghum roots release cyanogenic glucoside during decomposition (Putnam, 1983) which inhibits the growth of weeds and of some succeeding crops. However, whether such effects are truly allelopathic or simply the natural result of decomposition is still uncertain.

FUTURE RESEARCH

This brief review has shown that controlled experiments have been employed to good effect in understanding the responses of roots to the physical environment. However, predicting the magnitude of the response is still difficult because of the interactions of physical, chemical and biological factors. The work of Klepper *et al.* (1984), Gregory (1983) and Whiteley & Dexter (1983, 1984a, 1984b) have predictive value for crops and the ideas developed therein will no doubt also be applied to other plants in the future.

Other effects of the physical environment (notably effects of aeration) have only recently been measured satisfactorily so that it may be several years before models are available. Similarly the biological environment of roots is still being studied but few studies so far have predictive value (see the last paper in Atkinson *et al.* (1983) in relation to mycorrhiza). Until the ecology and effects of the physical environment on micro-organisms are quantified, their role in influencing plant growth will remain a matter for informed speculation.

REFERENCES

Abbas Al-Ani, M.K. & Hay, R.K.M. (1983). The influence of growing temperature on the growth and morphology of cereal seedling root systems. *Journal of Experimental Botany*, 34, 1720-1730.

Abbott, L.K. & Robson, A.D. (1982). The role of vesicular arbuscular mycorrhizal fungi in agriculture and the selection of fungi for inoculation. *Australian Journal of Agricultural Science*, 33, 389-408.

Alafifi, M.A. (1983). *Effects of Temperature on the Growth and Activity of Roots of Spring Wheat*. England: PhD Thesis, University of Reading.

Albertson, F.W. (1937). Ecology of mixed prairie in west central Kansas. *Ecological Monographs*, 7, 481-547.

Angus, J.F., Cunningham, R.B., Moncur, M.W. & Mackenzie, D.H. (1981). Phasic development in field crops 1. Thermal response in the seedling phase. *Field Crops Research*, 3, 365-378.

Arkin, G.F. & Taylor, H.M. (1981). *Modifying the Root Environment to Reduce Crop Stress*. American Society of Agricultural Engineers Monograph No. 4, Michigan, U.S.A.

Armenta-Soto, J., Chang, T.T., Loresto, G.C. & O'Toole, J.C. (1983). Genetic analysis of root characters in rice. *SABRAO Journal*, 15, 103-116.

Atkinson, D. (1983). The growth, activity and distribution of the fruit tree root system. *Plant and Soil*, 71, 23-35.

Atkinson, D., Bhat, K.K.S., Coutts, M.P., Mason, P.A. & Read, D.J. (1983). *Tree Root Systems and Their Mycorrhizas*. Nijhoff/Junk, The Hague.

Baker, C.K., Gallagher, J.N. & Monteith, J.L. (1980). Daylength change and leaf appearance in winter wheat. *Plant, Cell and Environment*, 3, 285-287.

Barley, K.P. (1963). Influence of soil strength on growth of roots. *Soil Science*, 96, 175-180.

Barley, K.P. (1970). The configuration of the root system in relation to nutrient uptake. *Advances in Agronomy*, 22, 159-201.

Barley, K.P. & Greacen, E.L. (1967). Mechanical resistance as a soil factor influencing the growth of roots and underground shoots. *Advances in Agronomy*, 19, 1-43.

Barraclough, P.B. (1984). The growth and activity of winter wheat roots in the field: root growth of high-yielding crops in relation to shoot growth. *Journal of Agricultural Science, Cambridge*, 103, 439-442.

Barraclough, P.B. & Leigh, R.A. (1984). The growth and activity of winter wheat roots in the field: the effect of sowing date and soil type on root growth of high-yielding crops. *Journal of Agricultural Science, Cambridge*, 103, 59-74.

Belford, R.K., Cannell, R.Q., Thomson, R.J. & Dennis, C.W. (1980). Effects of waterlogging at different stages of development on the growth and yield of peas (*Pisum sativum* L.). *Journal of Science of Food and Agriculture*, 31, 857-869.

Black, A.L. (1970). Soil water and soil temperature influences on dryland winter wheat. *Agronomy Journal*, 62, 797-801.

Blacklow, W.M. (1972). Influence of temperature on germination and elongation of the radicle and shoot of corn (*Zea mays* L.). *Crop Science Science*, 12, 647-650.

Brown, S.C. (1984). *Varietal variation in Root Growth, Yield and Water Use of Barley grown at two Sites in Northern Syria*. Report No.2 ODA Research Scheme R3691, Department of Soil Science, University of Reading.

Buwalda, J.G. & Goh, K.E. (1982). Host-fungus competition for carbon as a cause of growth depression in vesicular-arbuscular mycorrhizal ryegrass. *Soil Biology and Biochemistry*, 14, 103-106.

Cannell, R.Q. & Belford, R.K. (1980). Effects of waterlogging at different stages of development on the growth and yield of winter oilseed rape (*Brassica napus* L.). *Journal of Science of Food and Agriculture*, 31, 963-965.

Cannell, R.Q., Belford, R.K., Gales, K., Dennis, C.W. & Prew, R.D. (1980). Effects of waterlogging at different stages of development on the growth and yield of winter wheat. *Journal of Science of Food and Agriculture*, 31, 117-132.

Caradus, J.R. (1976). Structural variation of white clover root systems. *New Zealand Journal of Agricultural Research*, 20, 213-219.

Chapin, F.S. (1980). The mineral nutrition of wild plants. *Annual Review of Ecology and Systematics*, 11, 233-260.

Cooper, A.J. (1973). *Root Temperature and Plant Growth*. Research Review No. 4, Commonwealth Bureau of Horticulture and Plantation Crops, East Malling, Maidstone, Kent.

Cornish, P.S. (1982). Root development in seedlings of ryegrass (*Lolium perenne* L.) and phalaris (*Phalaris aquatica* L.) sown onto the soil surface. *Australian Journal of Agricultural Research*, 33, 665-677.

Coutts, M.P. & Philipson, J.J. (1978). Tolerance of tree roots to waterlogging II. Adaptation of Sitka spruce and Lodgepole pine to waterlogged soils. *New Phytologist*, 80, 71-77.

Cullen, P.W., Turner, A.K. & Wilson, J.H. (1972). The effect of irrigation depth on root growth of some pasture species. *Plant and Soil*, 345-352.

Feldman, L.J. (1984). Regulation of root development. *Annual Review of Plant Physiology*, 35, 223-242.

Fowkes, N.D. & Landsberg, J.J. (1981). Optimal root systems in terms of water uptake and movement. In *Mathematics and Plant Physiology* eds. D.A. Rose & D.A. Charles-Edwards, pp. 109-125. London: Academic Press.

Gallagher, J.N. (1979). Field studies of cereal leaf growth 1. Initiation and expansion in relation to temperature and ontogeny. *Journal of Experimental Botany*, 30, 625-636.

Garwood, E.A. (1967). Some effects of soil water conditions and soil temperature on the roots of grasses 1. The effect of irrigation on the weight of root material under various swards. *Journal of the British Grassland Society*, 22, 176-181.

Gerard, C.J., Sexton, P. & Shaw, G. (1982). Physical factors influencing soil strength and root growth. *Agronomy Journal*, 74, 875-879.

Gill, W.R. & Bolt, G.H. (1955). Pfeffer's studies of the root growth pressures exerted by plants. *Agronomy Journal*, 47, 166-168.

Goss, M.J. (1977). Effects of mechanical impedance on root growth in barley (*Hordeum vulgare* L.) 1. Effects on the elongation and branching of seminal root axes. *Journal of Experimental Botany*, 28, 96-111.

Greacen, E.L. & Oh, J.S. (1972). Physics of root growth. *Nature (New Biology)*, 235, 24-25.

Gregory, P.J. (1976). *The Growth and Activity of Wheat Root Systems.* England: PhD Thesis, University of Nottingham.

Gregory, P.J. (1983). Response to temperature in a stand of pearl millet (*Pennisetum typhoides* S. & H.) III Root development. *Journal of Experimental Botany*, 34, 744-756.

Gregory, P.J. (1986). Response to temperature in a stand of pearl millet (*Pennisetum typhoides* S. & H.) VIII Root growth. *Journal of Experimental Botany*, 37, 379-88.

Gregory, P.J., McGowan, M., Biscoe, P.V. & Hunter, B. (1978). Water relations of winter wheat I. Growth of the root system. *Journal of Agricultural Science, Cambridge*, 91, 91-102.

Grime, J.P. (1979). *Plant Strategies and Vegetation Processes*. New York: Wiley.

Hadley, P., Roberts, E.H., Summerfield, R.J. & Minchin, F.R. (1984). Effects of temperature and photoperiod on flowering on soyabean (*Glycine max* (L.) Merrill): a quantitative model. *Annals of Botany*, 53, 669-681.

Hendrickson, A.H. & Veihmeyer, F.J. (1931). Influence of dry soil on root extension. *Plant Physiology*, 6, 567-576.

Hewitt, J.S. & Dexter, A.R. (1984). The behaviour of roots encountering cracks in soil II. Development of a predictive model. *Plant and Soil*, 79, 11-28.

Hosier, P.E. & Eaton, T.E. (1980). The impact of vehicles on dune and grassland vegetation on a south-eastern North Carolina barrier beach. *Journal of Applied Ecology*, 17, 173-182.

Hurd, E.A. (1974). Phenotype and drought tolerance in wheat. *Agricultural Meteorology*, 14, 39-55.

Jackson, M.B. & Drew, M.C. (1984). Effects of flooding on growth and metabolism of herbaceous plants. In *Flooding and Plant Growth* ed. T.T. Kozlowski, pp. 47-127, London: Academic Press.

Kawase, M. (1981). Anatomical and morphological adaptation of plants to waterlogging. *Hortscience*, 16, 30-34.

Kirby, E.J.M., Appleyard, M. & Fellowes, G. (1982). Effect of sowing date on the temperature response of leaf emergence and leaf size in barley. *Plant, Cell and Environment*, 5, 477-484.

Klepper, B., Belford, R.K. & Rickman, R.W. (1984). Root and shoot development in winter wheat. *Agronomy Journal*, 76, 117-122.

Kordan, H.A. (1976). Adventitious root initiation and growth in relation to oxygen supply in germinating rice seedlings. *New Phytologist*, 76, 81-86.

Kutschera, L. (1960). *Wurzelatlas mitteleuropaischer Ackerunkrauter und Kulturpflanzen*. Frankfurt-am-Main: D.L.G. Verlags, Gmbh.

Locke, L.F. & Allen, J.A. (1924). Development of wheat plants from seminal roots. *Journal of American Society of Agronomy*, 16, 261-268.

McGowan, M., Wellings, S.R. & Fry, G.J. (1983). The structural improvement of damaged clay subsoils. *Journal of Soil Science*, 34, 233-248.

McIntosh, M.S. & Miller, D.A. (1980). Development of root-branching in three alfalfa cultivars. *Crop Science*, 20, 807-809.

Mengel, D.B. & Barber, S.A. (1974). Development and distribution of the corn root system under field conditions. *Agronomy Journal*, 66, 341-344.

Merrill, S.D. & Rawlins, S.L. (1979). Distribution and growth of sorghum roots in response to irrigation frequency. *Agronomy Journal*, 71, 738-745.

Nielsen, K.F. (1974). Roots and root temperature. In *The Plant Root and Its Environment*. ed. E.W. Carson, pp. 293-333. The University Press of Virginia, Charlottesville.

Nishiyama, I. (1976). Effects of temperature on the vegetative growth of rice plants. In *Climate and Rice*, pp. 159-185. Philippines: International Rice Research Institute, Los Banos.

van Noordwijk, M. (1983). Functional interpretation of root densities in the field for nutrient and water uptake. In *Root Ecology and its Practical Application*, International Symposium, Gumpenstein, 1982. pp. 207-226.

O'Brien, L. (1979). Genetic variability of root growth in wheat (*Triticum aestivum* L.). *Australian Journal of Agricultural Research*, 30, 587-595.

Ong, C.K. (1983a). Response to temperature in a stand of pearl millet (*Pennisetum typhoides* S. & H.) 1. Vegetative development. *Journal of Experimental Botany*, 34, 322-336.

Ong, C.K. (1983b). Response to temperature in a stand of pearl millet (*Pennisetum typhoides* S. & H.) 4 Extension of individual leaves. *Journal of Experimental Botany*, 34, 1731-1739.

Pang, P.C. & Paul, E.A. (1980). Effects of vesicular-arbuscular mycorrhiza on ^{14}C and ^{15}N distribution in nodulated faba beans. *Canadian Journal of Soil Science*, 60, 241-250.

Portas, C.A.M. & Taylor, H.M. (1976). Growth and survival of young plant roots in dry soil. *Soil Science*, 121, 170-175.

Putnam, A.R. (1983). Allelopathic chemicals: nature's herbicides in action. *Chemistry and Engineering News*, April 1983, 34-45.

Reid, C.P.P. & Bowen, G.D. (1979). Effects of soil moisture on V/A mycorrhizal formation and root development in *Medicago*. In *The soil-root interface*, eds. J.L. Harley & R. Scott Russell, pp 211-220. London: Academic Press.

Rice, E.L. (1984). *Allelopathy*. London: Academic Press.

Richards, R.A. & Passioura, J.B. (1981). Seminal root morphology and water use of wheat II. Genetic variation. *Crop Science*, 21, 253-255.

Robertson, B.M., Waines, J.G. & Gill, B.S. (1979). Genetic variability for seedling root numbers in wild and domesticated wheats. *Crop Science*, 19, 843-847.

Rogers, C.E., Unger, P.W. & Kreitner, G.L. (1984). Adventitious rooting in 'Hopi' sunflower: function and anatomy. *Agronomy Journal*, 76, 429-434.

Rowse, H.R. (1974). The effect of irrigation on the length, weight, and diameter of lettuce roots. *Plant and Soil*, 40, 381-391.

Sanders, F.E., Mosse, B. & Tinker, P.B. (1975). *Endomycorrhizas*. London: Academic Press.

Sanders, F.E. & Sheikh, N.A. (1983). The development of vesicular-arbuscular mycorrhizal infection in plant root systems. *Plant and Soil*, 71, 223-246.

Schaff, B.E. & Skogley, E.A. (1982). Diffusion of potassium, calcium, and magnesium in Bozeman silt loam as influenced by temperature and moisture. *Soil Science Society of America Journal*, 46, 521-524.

Sivakumar, M.V.K., Taylor, H.M. & Shaw, R.H. (1977). Top and root relations of field-grown soybeans. *Agronomy Journal*, 69, 470-473.

Snellgrove, R.C., Splittetoesser, W.E., Stribley, D.P. & Tinker, P.B. (1982). The distribution of carbon and the demand of the fungal symbiont in leek plants with vesicular-arbuscular mycorrhizas. *New Phytologist*, 92, 75-87.

Sprent, J.I. (1975). Adherance of sand particles to soybean roots under water stress. *New Photologist*, 74, 461-463.

Stone, J.A. & Taylor, H.M. (1983). Temperature and the development of the taproot and lateral roots of four indeterminate soybean cultivars. *Agronomy Journal*, 75, 613-618.

Taylor, H.M. & Arkin, G.F. (1981). Root zone modifications: fundamentals and alternatives. In *Modifying the Root Environment to Reduce Crop Stress* ed. G.F. Arkin & H.M. Taylor, pp. 3-17. American Society of Agricultural Engineers Monograph No. 4.

Taylor, H.M. & Klepper, B. (1978). The role of rooting characteristics in the supply of water to plants. *Advances in Agronomy*, 30, 99-128.

Taylor, H.M. & Ratcliffe, L.F. (1969a). Root growth pressures of cotton, peas, and peanuts. *Agronomy Journal*, 61, 398-402.

Taylor, H.M. & Ratcliffe, L.F. (1969b). Root elongation rates of cotton and peanuts as a function of soil strength and soil water content. *Soil Science*, 108, 113-119.

Tinker, P.B. (1984). The role of microorganisms in mediating and facilitating the uptake of plant nutrients from soil. *Plant and Soil*, 76, 77-91.

Troughton, A. (1980). Production of root axes and leaf elongation in perennial ryegrass in relation to dryness of the upper soil layer. *Journal of Agricultural Science, Cambridge*, 95, 533-538.

Weaver, J.E. (1926). *Root Development of Field Crops*. New York: McGraw-Hill.

Webster, R. & Beckett, P.H.T. (1972). Matric suction to which soils in South Central England drain. *Journal of Agricultural Science, Cambridge*, 78, 379-387.

Whiteley, G.M. & Dexter, A.R. (1981). The dependence of penetrometer pressure on penetrometer size. *Journal of Agricultural Engineering Research, 26, 467-476.*

Whiteley, G.M. & Dexter, A.R. (1983). Behaviour of roots in cracks between soil peds. *Plant and Soil*, 74, 153-162.

Whiteley, G.M. & Dexter, A.R. (1984a). Displacement of soil aggregates by elongating roots and emerging shoots of crop plants. *Plant and Soil*, 77, 131-140.

Whiteley, G.M. & Dexter, A.R. (1984b). The behaviour of roots encountering cracks in soil I. Experimental methods and results. *Plant and Soil*, 77, 141-149.

Wilson, A.J., Robards, A.W. & Goss, M.J. (1977). Effects of mechanical impedance on root growth in barley (*Hordeum vulgare* L.) II Effects on cell development in seminar roots. *Journal of Experimental Botany*, 28, 1216-1227.

Wilson, A.M., Hyder, D.N. & Briske, D.D. (1976). Drought resistance characteristics of blue grama seedlings. *Agronomy Journal*, 68, 479-484.

Yoshida, S. & Hasegawa, S. (1982). The rice root system: its development and function. In *Drought Resistance in Crops with emphasis on Rice*, pp 97-114. Philippines: International Rice Research Institute, Los Banos.

COMPETITION BETWEEN ROOT SYSTEMS IN NATURAL COMMUNITIES

M.M. Caldwell

INTRODUCTION

The distribution of plants in non-agricultural landscapes is moulded largely by the balance of competition between different plant species. Although competition for light is apparent in many communities such as forests, where plants are densely packed, much of the competition actually takes place below-ground in most ecosystems. The root systems of neighbouring plants compete for soil resources such as moisture and nutrients, yet, the manner in which these root systems interact and compete is neither obvious nor well studied. Competition below ground is known more by its manifestations than by its mechanisms. Even the most basic questions have received little attention. How much do roots of neighbouring plants overlap? How much competition takes place between individual root elements of neighbouring plants? How important is the domination of below-ground space for successful competition? This paper attempts to answer such questions using information drawn largely from experiments conducted in the field.

Competition among roots for resources can result if either roots of one plant deplete the soil resources more quickly than roots of another, or roots of the more successful competitor deplete resources to levels below which other plant roots can extract sufficient quantities for growth and survival (Tilman, 1982). Other interactions between plant root systems may, however, also be involved. For example, the colonization of roots by bacteria and fungi can be influenced by the presence of roots of other species (Christie, Newman & Campbell, 1978), and exudates from plant roots or decay products of dead roots may influence the growth or function of roots of other species (Rice, 1974). The possibilities for indirect interactions between roots seem almost endless and, indeed, many have been demonstrated in container-grown plants. Nevertheless, the significance of such indirect interactions in natural communities is still largely unknown. Interference is a generic term often used to indicate any modification of the plant environment by neighbouring plants. Competition by resource depletion is only a special case of interference. However, no matter how indirect the processes of interference may be, the end result is that one plant often has the advantage and, in the end, gains more resources. Competition, as used in this paper, will refer to any situation where soil resources are in limited supply and one neighbour acquires more of the resources than the other, irrespective of the mechanism.

EVIDENCE OF BELOW-GROUND COMPETITION

There is a considerable literature on the manifestations of competition between plants. The influence of competition is clearly indicated by the improved growth, seed yield, and water status of plants that occur when neighbouring plants are removed (Robberecht, Mahall & Nobel, 1983; Ehleringer, 1984). The relationship between plant size and distance from neighbouring plants also provides indirect evidence of the degree of competitive interference between plants (Pielou, 1962; Barbour, 1969). For example, there is a distinct relationship between the size of individual plants in a monospecific stand of the Sonoran Desert shrub *Encelia farinosa* when size is plotted against the distance between neighbouring plants (Fig. 1). Such a relation hardly proves competition, but at least infers it. Implicit in this inference is that resources are limiting, that resources are sufficiently uniform in their distribution (i.e., resources and space are well correlated), that plant size and resource acquisition are proportional, and that resources not used by one plant will be exploited by neighbours.

The study of plant competition has also involved observation of the performance of plants transplanted into different communities, and the performance of species when sown as either mixtures or monocultures. Much is now known about which species are most likely to compete favourably in different environments. However, the specific characteristics that cause plants to be more competitive in certain environments, particularly below-ground, are less well understood.

In a desert, much of the competition might be expected to take place below ground since water is likely to be the most limiting resource in that environment; competition for light is less likely. Even in situations where plant shoots are crowded and compete for light, root competition can be even more severe. A classic experiment by Donald (1958) demonstrated this vividly. *Lolium perenne* and *Phalaris tuberosa* were grown together in pots

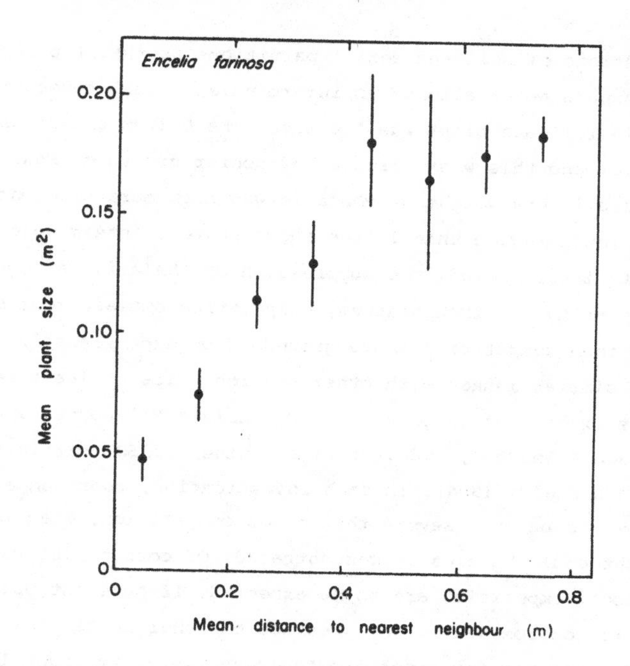

Figure 1. The relationship between the average size and the average distance between nearest neighbours in a monospecific stand of *Encelia farinosa* in the Sonoran Desert of Arizona. Adapted from Ehleringer (1984).

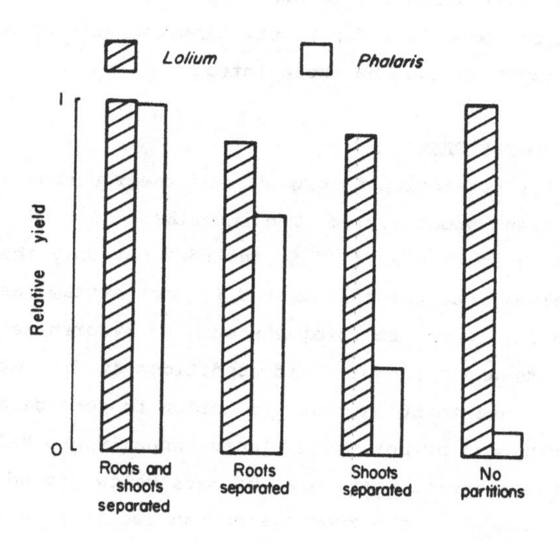

Figure 2. Relative changes in yield of *Phalaris tuberosa* and *Lolium perenne* grown in pots with different combinations of aerial and root partitions. From data of Donald (1958).

with various combinations of soil and aerial partitions so that the roots or shoots could be separated or allowed to intermingle. In all cases, the soil volume available for each plant was the same. The *Lolium* grass was competitively superior and this was expressed by poorer growth of *Phalaris* (Fig. 2). The decrease in dry weight of *Phalaris* was much more pronounced if the root systems intermingled than if the shoot systems intermingled. If both shoots and roots intermingled, the suppression of *Phalaris* was most pronounced. Thus, under these circumstances, competition between root systems was more pronounced than competition above ground. Such experiments have been performed in a similar manner with other species pairs at least seven times since Donald's experiment (Aspinall, 1960; Idris & Milthorpe, 1966; Snaydon, 1971; Remison & Snaydon, 1980; Scott & Lowther, 1980; Martin & Snaydon 1982; Martin & Field, 1984). In each investigation, root competition was always shown to be more severe than shoot competition, even though competition for light could be clearly demonstrated. Of course, interactions between root and shoot competition are to be expected. If nutrient uptake by the root system is compromised by the presence of other roots, this may limit the supply of nutrients for shoot growth. Conversely, if light interception by the shoot system is made less effective due to shading by competitors, less photosynthate will be available for root growth. Although extrapolation from such controlled studies to plant communities at large is hardly warranted, at least the potential for significant competition among root systems in natural ecosystems must be appreciated.

INVESTMENT IN ROOT SYSTEMS

Roots are usually heterotrophic organs, and their carbon economy must be subsidized by the plant shoot system. The investment in root systems in natural communities can be considerable. This includes not only the production of roots themselves, but carbon invested in root system respiration, exudates, and carbon commitment to symbionts such as mycorrhizal fungi. Quantification of these costs under actual field conditions is difficult (to say the least). If only the investment in root production is considered, more than half of the tissues produced by perennial plants annually are below ground (Fig. 3). Even if the proportion of total biomass below ground is small, the annual rate of renewal of the root system can result in a large investment in below-ground production. For example, the renewal rate of the fine root system in a pine plantation system in Sweden has been estimated to be in excess of 600% each year (Agren *et al.*, 1980; Persson, 1983). In such diverse ecosystems as arctic tundra, deserts, and deciduous forest, the proportion of primary production directed below ground substantially exceeds that above ground. Although the methods employed to obtain the data shown in Fig. 3 vary considerably and the errors incurred are often sizeable, the magnitude of below ground production relative to that above ground clearly emerges as a significant investment. When the unmeasured costs such as root respiration and exudates are added to these production costs, root systems clearly represent a large drain on the carbon economy of plants.

How much of this large investment in root systems is necessary simply to procure soil moisture and nutrients and how much is needed to compete effectively for soil resources with neighbouring plants? These questions are intriguing but presently unanswerable because so little is known about the way in which roots compete for soil resources. There are some indications, however, that plants may produce a larger root system than is needed simply for uptake of soil resources and that this extra investment

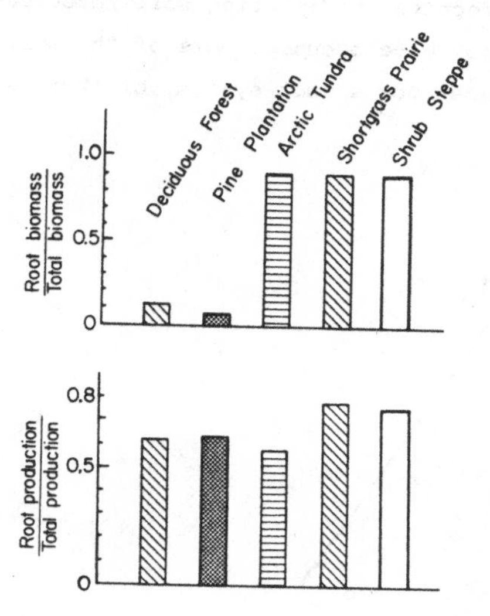

Figure 3. The proportion of plant biomass in root systems of different ecosystems dominated by perennial plants and the proportion of total annual production in the root systems for these ecosystems. Data from a deciduous forest dominated by *Liriodendron tulipifera* (Reichle *et al.*, 1973), a 14-year *Pinus sylvestris* plantation (Persson, 1983 and Agren *et al.*, 1980), an arctic tundra in northern Alaska (Shaver & Billings, 1975), an ungrazed shortgrass prairie in eastern Colorado (Sims & Singh, 1978), and a cool desert shrub steppe dominated by *Atriplex confertifolia* (Caldwell *et al.*, 1977).

may serve primarily for increased competitive effectiveness.

The first indication comes from studies with different genetic lines of perennial ryegrass, *Lolium perenne*, bred in Wageningen, The Netherlands. These lines differed consistently in root/shoot ratios and in competitiveness with other *Lolium* clones and with a highly competitive grass, *Elytrigia repens* (Fig. 4) (Baan Hofman & Ennik, 1980 and 1982; Ennik & Baan Hofman, 1983). The most competitive *Lolium* clones were consistently those with greater root mass and smaller shoot/root ratios. In monoculture, the shoot production of the more competitive clones was never more, and often less, than the less competitive clones. However, in competition with other *Lolium* clones and with *Elytrigia*, production of the more competitive clones increased at the expense of the competitors. These differences in competitiveness were also evident when nutrients and water were liberally supplied in these experiments (Baan Hofman & Ennik, 1982). Exactly why the more competitive clone had the advantage is not clear. Their more massive root systems may have been more effective in depleting soil resources, at least in localized zones, or they may have occupied more of the soil volume to the partial exclusion of the competitor's root system, or there may have

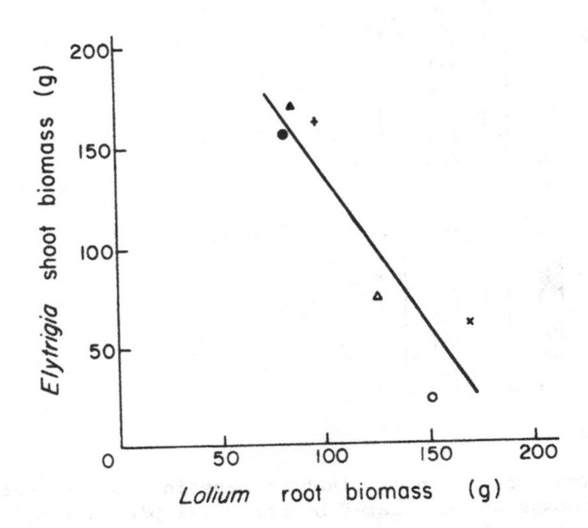

Figure 4. Shoot biomass of *Elytrigia repens* produced when growing in competition with different clones of *Lolium perenne* compared to the root biomass produced by the different *Lolium* clones grown in monoculture. The different symbols represent values for the different clones. Adapted from Ennik & Baan Hofman (1983).

been still other interactions apart from resource competition. Unfortunately, the rooting patterns and degree of intermingling of the root systems were not investigated in these studies.

The *Lolium* clone experiments indicate that investment in greater root mass is related to enhanced competitive effectiveness and that this greater root mass was not necessarily advantageous to the grasses growing in monoculture. In the absence of competition, one might conclude, then, that plants have a larger root system than is needed.

Theoretical estimates of the rooting density (length of root per volume of soil) necessary to acquire moisture and nutrients have been made (van Noordwijk, 1983). These calculations employed models of water and nutrient uptake by root systems taking account of root dimensions and absorptive capacities, and the characteristics of water and nutrient movement under particular soil conditions. Rooting densities in a 20 cm layer of soil were calculated for adequate uptake of water, nitrate and phosphate to be 0.3 to 5 cm (root) cm^{-3} (soil), 0.1 to 1 cm^{-2} and 1 to 10 cm^{-2}, respectively. The range of rooting densities for each resource reflects the range of soil conditions and assumptions employed in the models. As the abundance of a soil resource decreases or the resistance to its movement in the soil or into the root increases, the required rooting density for adequate supply increases. The higher rooting density required for phosphate uptake compared to nitrate or water reflects the low diffusivity of phosphate in soil, and, thus, the need for a dense network of roots to effect the necessary supply. Rooting densities of about 10 cm^{-2} are not uncommon in the upper soil layers and values for cereals greater than 15 cm^{-2} have been reported in some soils (Greacen, Ponsana & Barley, 1976).

It is well known that when plants are growing in monoculture, they can alter the proportion of roots to shoots. If soils are deficient in moisture or nutrients, many plants apportion more biomass to the root system (Nye & Tinker, 1977). There is little evidence to indicate whether plants in competition exhibit a similar flexibility and apportion more biomass to root systems although there is some indication of this in the studies of Ennik & Baan Hofman (1983). If this occurred generally, it would be further circumstantial evidence for the hypothesis that plants invest in roots systems to compete effectively.

The theoretical calculations as well as the empirical, circumstantial evidence from the *Lolium* clones suggest that plants may, indeed, invest more in root systems than would be necessary in the absence of competition to acquire soil moisture and nutrients, at least under some circumstances. This theme will recur.

THE NATURE OF BELOW-GROUND COMPETITION

Do root systems compete for resources primarily by occupying a soil volume to the complete or partial exclusion of neighbouring root systems? If so, soil moisture and nutrients would be merely the spoils of conquering a certain space below ground. If neighbouring root systems intermingle and specific soil volumes are not so clearly dominated by one plant or another, is competition a phenomenon taking place among individual roots in close proximity with one another? Some clues to such questions can be gleaned from descriptions of root system distributions or, in some cases, from soil moisture or nutrient utilization patterns. If little overlap of neighbouring root systems is apparent, or if moisture or nutrients are drawn from distinctly different soil zones by neighbouring plants, there will be little competition between individual roots of neighbouring plants, and resources will be freely used by the plant occupying a certain soil space. Surprisingly, there is little information relating specifically to the question of overlap between neighbouring root systems and what evidence that exists is equivocal.

Root systems of strikingly different morphology such as those with deep penetrating tap roots, shallow fibrous roots, or horizontally extending rhizomes, may overlap very little and these different root forms can avoid competition. The issue of overlap and competition then relates primarily to root systems of similar form.

Extensive excavation of root systems has been conducted in orchards. Some studies reveal surprisingly little overlap of neighbouring root systems, so much so that terms such as "antagonism" between root systems have been used (Rogers & Head, 1969). The degree of overlap appears to depend on the species of tree. However, even for the same species, Atkinson, Naylor & Coldrick (1976) found that both the degree of overlap and the general morphology of the root systems may differ (Fig. 5). In denser plantings, the degree of overlap increased considerably and in some cases trees were sharing soil space with trees two rows away. The root systems of widely spaced trees were almost discrete and only occasionally intermingled at the periphery whereas those at narrow spacings had a larger number of the major roots growing downward rather than spreading horizontally. The crowded trees also had a greater proportion of finer roots. Thus, the behaviour of root systems can be expected to change depending on the proximity of neighbours and it is difficult to predict the extent of root system overlap, even for these cultivated tree species.

In natural plant communities, spacings between plants are usually very irregular and several species may be neighbours; this constitutes a more complicated system than an orchard. The least overlap of root systems might be expected with widely spaced plants in deserts. The creosote bush of North America, *Larrea tridentata*, has been the object of much speculation about the nature of its competition, both intra- and inter-specific. This species is widely distributed throughout the hot deserts of North America where competition for moisture would be expected. Much attention has been

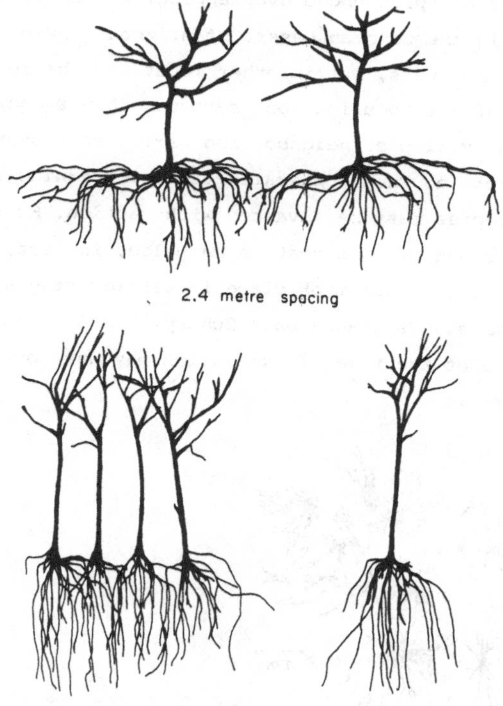

Figure 5. Effect of different spacings on the morphology of root systems of apple trees excavated from an orchard in England. A single tree removed at the denser spacing is also shown in the lower right portion of the figure. Drawn from photographs of Atkinson *et al*. (1976).

focussed on the dispersion patterns (whether clumped, random or evenly spaced) of this shrub and what this might imply regarding the degree of intra- or inter-specific competition. A discussion of these dispersion patterns is of less interest in this review than the evidence for overlap of root systems.

Figure 6 shows an excavation of a root system at a site in southern New Mexico where the shrubs were not evenly spaced. Despite the uneven spacing there appeared to be little overlap of neighbouring systems and nearly all of the below-ground space in the rooting depth was filled with roots. The very fine roots were not measured in this excavation, and they might have overlapped. Somehow the roots of neighbouring shrubs adjusted their growth patterns such that the below-ground space was fully exploited and yet there was not pronounced overlapping. The early root mappings of Cannon (1911) in the Sonoran Desert of Arizona showed a similar pattern for *Larrea* in some habitats, but in other locations, he found considerable intertwining of neighbouring root systems. Chew & Chew (1965) also reported considerable overlap of neighbouring *Larrea* root systems in their excavations in California. Thus, generalizations concerning root system overlap for this species must be advanced with caution. Furthermore, the distributions of plants and of root systems may also, in part, reflect interactions between neighbours which took place in earlier stages of development of these *Larrea* stands (MacMahon & Schimpf, 1981). Root systems which appear to be discrete at the time of excavation may have overlapped considerably at an earlier time.

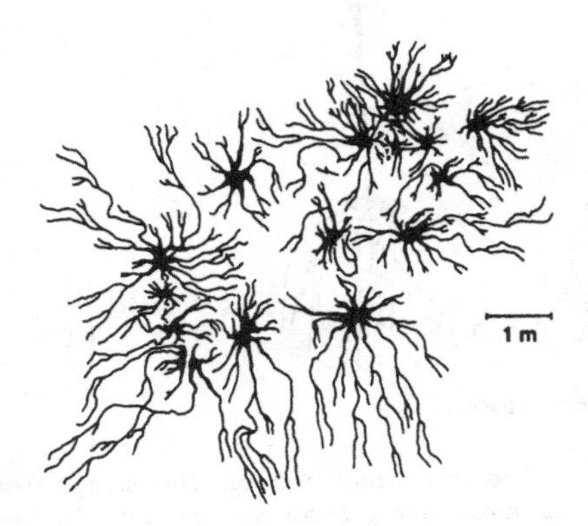

Figure 6. Horizontal root distributions of *Larrea tridentata* (Redrawn from Ludwig, 1977).

An indirect assessment of moisture competition by *Larrea* was reported in some experiments by Fonteyn & Mahall (1981). This study used *Larrea* and a smaller shrub, *Ambrosia dumosa*, which occur together in the Mojave Desert in southern California. Using circular, 100 m^2 plots with either an *Ambrosia* or a *Larrea* in the centre, various combinations of shrubs were removed to evaluate intra- and interspecific competition. They measured plant water potential of the central shrub just before dawn when the potential should be close to equilibrium with soil water potential. Fig. 7 shows the patterns of removal and the course of pre-dawn water potential of the central *Ambrosia* plants as the soil became depleted of moisture. Water potentials of central *Ambrosia* plants in plots where all other shrubs were removed differed from those of *Ambrosia* in the control plots when pre-dawn water potentials were very negative (-3 to -5 MPa). Removal of other *Ambrosia* alone did not significantly increase the water potential but removal of *Larrea* alone resulted in less negative plant water potentials in central *Ambrosia* plants. Where *Larrea* was the central plant, similar results were obtained although the differences were not as large. These results were reasonably consistent over three long drying cycles in an 18-month period.

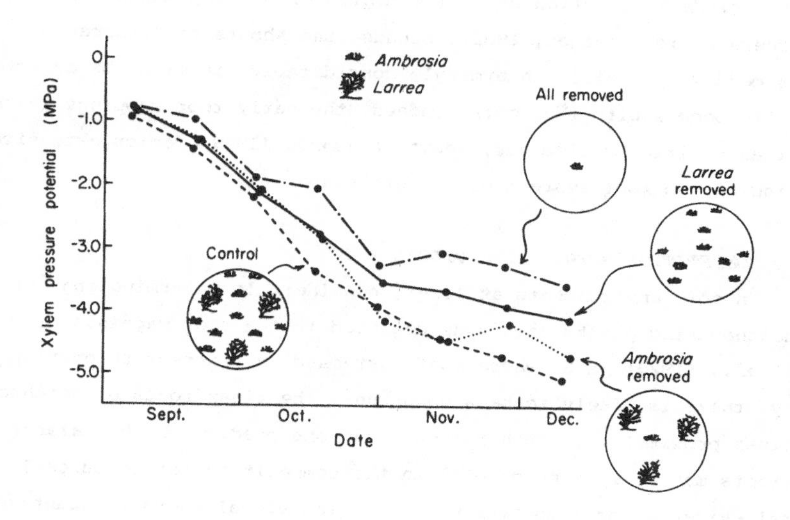

Figure 7. Changes in pre-dawn xylem water potentials of central shrubs of *Ambrosia dumosa* during the autumn of 1977 following selective removal of neighbouring shrubs early in 1977. The patterns of shrub removal in the 100 m^2 circular plots are illustrated in this figure. Adapted from Fonteyn & Mahall (1981).

Thus, near the end of a drying cycle, the central shrubs were tapping somewhat moister soil when shrubs of the other species were removed. However, at these low water potentials, the changes represent small differences in volumetric soil water content. Because removal of shrubs of the same species usually resulted in insignificant changes in plant water potential, the authors concluded that intra-specific competition among either *Ambrosia* or *Larrea* shrubs was much less pronounced than interspecific competition between *Ambrosia* and *Larrea*.

One might be tempted to speculate from these results that root systems of the central plants overlapped much less with those of the same species than they did with root systems of the other species. However, in the several months following shrub removals in the winter of 1977, considerable changes in the below-ground space occupied by the centre shrub may have occurred as soil space became available following the death of the root systems of those shrubs that had been removed. Thus, these intriguing results still do not shed light on how these root systems might be competing.

Even less attention has been directed toward root systems of herbaceous plants and even some of the most elaborate field descriptions of them do not include information about the degree of overlap among neighbours (e.g. Kutschera & Lichtenegger, 1982). Because the shoots of herbaceous plants in a meadow or prairie intermingle considerably, it might be expected that this also occurs with the roots. Indeed, the early root mappings in the North American prairie by Clements, Weaver & Hanson (1929) depict extensive intermingling of the root systems of different species.

COMPETITION BETWEEN INDIVIDUAL ROOTS

In some environments at least, considerable intermingling between roots of neighbouring plants occurs as depicted in the root mappings of Clements *et al.*, (1929). Even where root systems do not appear to overlap extensively, there is likely to be a zone where the finer roots of neighbours occur in close proximity. In such regions, can one predict which characteristics of roots may be most beneficial in the competition for resources? Experimental evidence for competition between individual roots is scant and one can only surmise which characteristics of roots systems would be of particular importance for competitive effectiveness.

Assuming that competition between individual roots is mediated by resource depletion, the root characteristics contributing to competitiveness might well be those that provide for rapid resource uptake in the absence of competition. Rather elaborate mathematical models of moisture and nutrient uptake have been developed which provide a means of evaluating the importance of physiological and morphological traits of individual roots (e.g., Cushman, 1979). Such models consider the movement of nutrients or water through a cylinder of soil to the root and take account of soil factors and uptake kinetics for different concentrations of the resource at the root surface. These models are largely mechanistic, but they depend on many parameters and assumptions that cannot be known exactly. Nevertheless, they give insight to the relative importance of different root characteristics. Generally, morphological parameters, especially rooting density (and hence the average diameter of the root cylinder around each root, since a greater rooting density results in less soil between roots), have a greater influence on uptake rates than the physiological absorption capacities of the root itself (Clarkson, 1985; Caldwell & Richards, 1986). This is mainly because resource movement through the soil is usually more limiting than absorption by the root itself. This depends on the resource in question and its mobility in a particular soil. As indicated earlier, a greater rooting density is particularly beneficial for efficient uptake of less mobile resources such as phosphate.

Because rooting density appears advantageous for the acquisition of soil resources in theoretical assessments, it might be assumed that competitive effectiveness for soil resources would also be substantially increased by a high rooting density. Circumstantial evidence for this assertion is provided by the association of greater competitive effectiveness with greater root mass of *Lolium* clones as cited earlier (Fig. 4). A comparison of two species of *Agropyron* tussock grasses that differ strikingly in their competitive effectiveness with a shrub, *Artemisia tridentata*, also provides some circumstantial evidence that greater rooting density is an advantage in competition. Similar-sized tussocks of these grasses have the same root mass, but one grass has much thinner roots and, therefore, greater rooting density for the biomass invested in the root system (Caldwell & Richards, 1986). The *Agropyron* species with the greater rooting density is more efficient in soil moisture extraction when in monoculture as well as being more competitive when grown with the shrub.

Although rooting density appears to be important in the models and in these two pieces of circumstantial evidence, other root characteristics may also have a large bearing on competitive effectiveness. Proton or phosphatase secretions that facilitate nutrient uptake, root hair characteristics, mycorrhizal associations, or simply flexibility in altering root growth patterns or physiological characteristics may, under certain circumstances, be of particular importance in gaining some competitive advantage. These, however, have not been well explored theoretically or experimentally in the context of competition. The flexibility of root system behaviour is potentially very important. Soil moisture and nutrients are not uniformly distributed spatially or temporally. Thus, the ability of root systems to change growth patterns and relocate zones of active absorption, even on a small scale, may be of particular advantage. The presence of root systems of other species also may result in shifts in active zones of absorption. This was vividly shown in phosphorus uptake studies of Goodman & Collison (1982). Using ^{32}P they demonstrated that the major zones of phosphorus uptake by *Lolium perenne* were deeper in the presence of clover than when the grass was growing in monoculture. Whether there was a major adjustment in root distributions or simply a change in activity of the roots at different depths is not known because no root excavations were conducted. Nevertheless, such changes of absorption zones have an important bearing on the nature of below-ground competition. Sometimes, competition between individual roots of neighbouring species may be greatly reduced if root systems shift locations of activity.

Apart from the question of root systems traits that may contribute to competitive effectiveness, models of moisture and nutrient absorption also provide some insight into the intensity of competition between individual roots for different soil resources as a function of soil conditions and the relative distance between roots. In the soil cylinder surrounding each root, soil moisture and nutrients may become depleted toward the root. The degree of depletion is more pronounced as the rate of extraction by the root increases relative to the mobility of the resource (Nye & Tinker, 1977). For an immobile resource such as phosphate, a marked depletion immediately near an active root is almost always to be expected. Further more, unless roots are very close to one another (i.e., within a few mm) these depletion zones do not overlap and, therefore, there is little competition between individual roots. If the roots are close enough for the depletion zones to overlap, competition is expected to be very intense. Conversely, the depletion zones for more mobile nutrients are less pronounced and extend further from the root so that the probability of overlap with depletion zones of neighbouring roots is greater, but the competition is less intense. Soil water potential also influences the development of depletion zones for both soil water and nutrients. The overlap of depletion zones is also complicated by the presence of root hairs and mycorrhizae that can extend the effective depletion zone for nutrients such as phosphate. Furthermore, as roots grow and thus change their location of absorption zones in the soil, this pattern necessarily changes. The use of theoretical models in assessing root competition is treated in more detail by Caldwell & Richards, (1986).

Experimental evidence demonstrating competition between roots of neighbouring plants in a very confined soil space comes from phosphorus isotope studies conducted in field plots (Fig. 8). Using two radioactive isotopes of phosphorus inserted halfway between a shrub and each of two neighbouring species of tussock grass, it was possible to compare the results of competition for phosphorus by the shrub with that of the two grass species. The soil was calcareous and the added phosphate was rapidly adsorbed. Active, mycorrhizal roots of both the grass and shrub were present in both interspaces where competition for the labelled phosphate occurred. Of the total radioactive phosphate absorbed by the shrub, six times as much came from

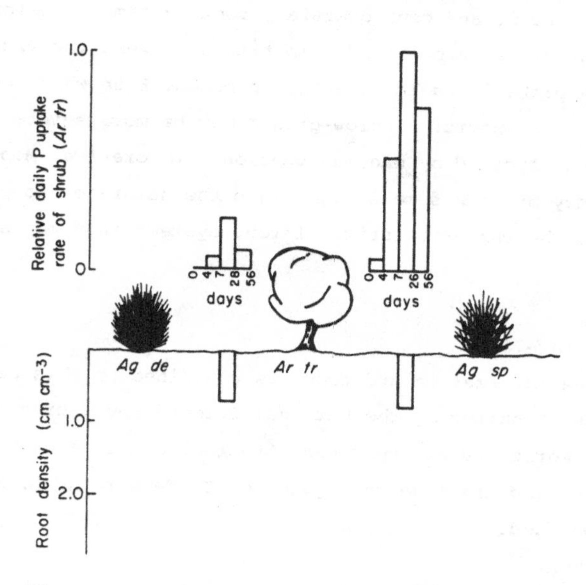

Figure 8. Relative rate of phosphorus isotope uptake by *Artemisia tridentata* (*Ar tr*) from labelled interspaces shared with *Agropyron spicatum* (*Ag sp*) and *Agropyron desertorum* (*Ag de*) at different times following placement of the phosphorus isotopes. Also shown is the root density of *Artemisia tridentata* in the centre of the interspaces between the shrub and each of its grass neighbours where the isotope labelling took place. Adapted from Caldwell *et al.* (1985).

the interspace with the weaker grass competitor as from the interspace shared with the stronger grass competitor even though the root length and mycorrhizal associations in both interspaces were identical. The ability of the stronger grass competitor to deplete resources at the expense of the shrub appears to be a prime factor in this differential uptake of phosphorus isotopes.

CONCLUDING REMARKS

Plants invest heavily in root systems, partly to compete effectively for below-ground resources. An investment in increased root length per unit volume of soil appears to be particularly important. In the last example cited, the presence of an effective competitor reduced the resource gain by the shrub relative to that removed in the presence of a weaker competitor even though the shrub was apparently investing the same in root systems. Thus, the presence of competing roots can increase the cost of root system investment in order to obtain the same quantity of resource. Although information is scant and controversial, root systems of neighbouring plants may frequently intermingle and competition between individual roots of neighbours is probably an important phenomenon. Even when shoots are competing for light, competition below-ground may be more severe in many instances. Although limited by many assumptions, theoretical models of root system activity may provide some insight into the nature of competition for soil resources and the characteristics of root systems that may enhance competitive potential.

ACKNOWLEDGEMENTS

Some of the information and concepts contained in this paper resulted from research supported by the National Science Foundation (DEB 8207171) and the Utah Agricultural Experiment Station. Comments by D. Eissenstat and S. Flint and drafting of figures by C. Warner and M. Mazurski are gratefully acknowledged.

REFERENCES

Agren, G.L., Axelsson, B., Flower-Ellis, J.G.K., Linder, S., Persson, H., Staaf, H. & Troeng, E. (1980). Annual carbon budget for a young Scots pine, In *Structure and Function of Northern Coniferous Forests: An Ecosystem Study*, ed. T. Persson, pp.307-13. Stockholm: Ecological Bulletin.

Aspinall, D. (1960). An analysis of competition between barley and white persicaria. 2. Factors determining the course of competition. *Annals of Applied Biology*, 48, 637-54.

Atkinson, D., Naylor, D. & Coldrick, G.A. (1976). The effect of tree spacing on the apple root system. *Horticultural Research*, 16, 89-105.

Baan Hofman, T. & Ennik, G.C. (1980). Investigation into plant characters affecting the competitive ability of perennial ryegrass (*Lolium perenne* L.). *Netherlands Journal of Agricultural Science*, 28, 97-109.

Baan Hofman, T. & Ennik, G.C. (1982). The effect of root mass of perennial ryegrass (*Lolium perenne* L.) on the competitive ability with respect to couchgrass (*Elytrigia repens* (L.) Desv.). *Netherlands Journal of Agricultural Science*, 30, 275-83.

Barbour, M.G. (1969). Age and space distribution of the desert shrub *Larrea divaricata*. *Ecology* 50, 679-85.

Caldwell, M.M., White, R.S., Moore, R.T. & Camp, L.B. (1977). Carbon balance, productivity, and water use of cold-winter desert shrub communities dominated by C_3 and C_4 species. *Oecologia* 29, 275-300.

Caldwell, M.M., Eissenstat, D.M., Richards, J.H. & Allen, M.F. (1985). Competition for phosphorus: Differential uptake from dual-isotope-labelled soil interspaces between shrub and grass. *Science*, 229, 384-6.

Caldwell, M.M. & Richards, J.H. (1986). Competing root systems: Morphology and models of absorption. In *On the Economy of Plant Form and Function*, ed. T.J. Givnish, pp 251-73. Cambridge: Cambridge University Press.

Cannon, W.A. (1911). *The Root Habits of Desert Plants*. Washington: Carnegie Institute Publication 131.

Chew, R.M. & Chew, A.E. (1965). The primary productivity of a desert shrub (*Larrea tridentata*) community. *Ecological Monographs*, 35, 355-75.

Christie, P., Newman, E.I. & Campbell, R. (1978). The influence of neighbouring grassland plants on each others' endomycorrhizae and root-surface microorganisms. *Soil Biology and Biochemistry*, 10, 521-7.

Clarkson, D.T. (1985). Factors affecting mineral nutrient acquisition by plants. *Annual Review of Plant Physiology*, 36, 77-115.

Clements, F.E., Weaver, J.E. & Hanson, H.C. (1929). *Plant Competition: An Analysis of Community Function*. Washington: Carnegie Institute Publication 398.

Cushman, J.H. (1979). An analytical solution to solute transport near root surfaces for low initial concentrations: I. Equation development. *Soil Science Society of America Journal*, 43, 1087-90.

Donald, C.M. (1958). The interaction of competition for light and nutrients. *Australian Journal of Agricultural Research*, 9, 421-35.

Ehleringer, J.R. (1984). Intraspecific competitive effects on water relations, growth and reproduction in *Encelia farinosa*. *Oecologia* 63, 153-8.

Ennik, G.C. & Baan Hofman, T. (1983). Variation in the root mass of ryegrass types and its ecological consequences. *Netherlands Journal of Agricultural Science*, 31, 325-34.

Fonteyn, P.J. & Mahall, B.E. (1981). An experimental analysis of structure in a desert plant community. *Journal of Ecology*, 69, 883-96.

Goodman, P.J. & Collison, M. (1982). Varietal differences in uptake of ^{32}P labelled phosphate in clover plus ryegrass swards and monocultures. *Annals of Applied Biology*, 100, 559-65.

Greacen, E.L., Ponsana, P., & Barley, K.P. (1976). Resistance to water flow in the roots of cereals. In *Water and Plant Life*, ed. O.L. Lange, L.Kappen & E.-D. Schulze, pp. 86-100. Heidelberg: Springer-Verlag.

Idris, H. & Milthorpe, F.L. (1966). Light and nutrient supplies in the competition between barley and charlock. *Oecologia Plantarum*, 1, 143-64.

Kutschera, L. & Lichtenegger, E. (1982). *Wurzelatlas mitteleuropaeischer Gruendlandpflanzen. Bd. 1. Monocotyledoneae*. Stuttgart: Gustav Fischer.

Ludwig, J.A. (1977). Distributional adaptations of root systems in desert environments. In *The Belowground Ecosystem: A Synthesis of Plant-Associated Processes*, ed. J.K. Marshall, pp. 85-91. Range Science Department Scientific Series No. 26. Fort Collins: Colorado State University.

Martin, M.P.L.D. & Snaydon, R.W. (1982). Root and shoot interactions between barley and field beans when intercropped. *Journal of Applied Ecology*, 19, 263-72.

Martin, M.P.L.D. & Field, R.J. (1984). The nature of competition between perennial ryegrass and white clover. *Grass and Forage Science*, 39, 247-53.

MacMahon, J.A. & Schimpf, D.J. (1981). Water in desert ecosystems. In *Water as a Factor in the Biology of North American Desert Plants*, US/IBP Synthesis Series, vol. 11, ed. D.D. Evans & J.L. Thames, pp 114-71. Stroudsburg, PA: Dowden, Hutchinson & Ross.

Noordwijk, M. van, (1983). Functional interpretation of root densities in the field for nutrient and water uptake. In *Wurzeloekologie und ihre Nutzanwendung/Root Ecology and its Practical Application*, International Symposium Gumpenstein, 1982. pp 207-26. Austria: Bundesanstalt Gumpenstein, Irdning.

Nye, P.H. & Tinker, P.B. (1977). *Solute Movement in the Soil-Root System*. Oxford: Blackwell Scientific Press.

Persson, H. (1983). The importance of fine roots in boreal forests. In *Wurzeloekologie und ihre Nutzanwendung/Root Ecology and its Practical Application*, International Symposium Gumpenstein, 1982, pp 505-608. Austria: Bundesanstalt Gumpenstein, Irdning.

Pielou, E.C. (1962). The use of plant-to-neighbour distances for detection of competition. *Journal of Ecology* 50, 357-67.

Reichle, D.E., Dinger, B.E., Edwards, N.T., Harris, W.F. & Sollins, P. (1973). Carbon flow and storage in a forest ecosystem. In *Carbon and the Biosphere*. ed., G.M. Woodwell & E.V. Pecan, pp 345-65. United States Atomic Energy Commission.

Remison, S.U. & Snaydon, R.W. (1980). A comparison of root competition and shoot competition between *Dactylis glomerata* and *Holcus lanatus*. *Grass and Forage Science*, 35, 183-7.

Rice, E.L. (1974). *Allelopathy*, New York: Academic Press.

Robberecht, R., Mahall, B.E. & Nobel, P.S. (1983). Experimental removal of intraspecific competitors-effects on water relations and productivity of a desert bunchgrass, *Hilaria rigida*. *Oecologia*, 60, 21-4.

Rogers, W.S. & Head, G.C. (1969). Factors affecting the distribution and growth of roots of perennial woody species. In *Root Growth*, Proceedings of the Fifteenth Easter School in Agricultural Science, University of Nottingham, 1968, ed. W.J. Whittington, London; Butterworths, pp 280-95.

Scott, R.S. & Lowther, W.L. (1980). Competition between white clover 'Grasslands Huia' and *Lotus pedunculatus* 'Grassland Maku'. I. Shoot and root competition. *New Zealand Journal of Agricultural Research*, 23, 501-7.

Shaver, G.R. & Billings, W.D. (1975). Root production and root turnover in a wet tundra ecosystem. Barrow, Alaska. *Ecology*, 56, 401-9.

Sims, P.L. & Singh, J.S. (1978). The structure and function of ten western North American grasslands. II. Intraseasonal dynamics in primary producer compartments. *Journal of Ecology* 66, 547-72.

Snaydon, R.W. (1971). An analysis of competition between plants of *Trifolium repens* L. populations collected from contrasting soils. *Journal of Applied Ecology* 8, 687-97.

Tilman, D. (1982). *Resource Competition and Community Structure*, Princeton, N.J: Princeton University Press.

RESPONSE OF ROOTS TO THE PHYSICAL ENVIRONMENT: GOALS FOR FUTURE RESEARCH

P.J.C. Kuiper

INTRODUCTION

It is a rather ambitious task to summarize the highlights of the papers presented at the symposium in Bangor and reported in this book. Several authors comment on the bewildering complexity of the specific root function which they discuss, from the cellular level of organization up to the level of root systems in plant communities.

A certain imbalance is evident in our understanding of the relation between the structure and function of root systems and this becomes even more evident when we try to understand structure and functioning of roots in heterogeneous soil environments instead of simplified water culture systems.

Nevertheless an attempt will be made to indicate goals for future research and the effects of the physical environment on plant roots. For convenience, our discussion will focus on the various levels of organization of the plant root system. The topics presented are a personal selection.

LEVELS OF ORGANIZATION

The cellular level

Plant roots are heterogeneous organs containing many cell types and tissues. The cells, in turn, are involved in a variety of functions including water absorption; uptake of ions and further translocation to the xylem vessels; re-absorption of ions from the xylem in older parts of the roots; extrusion of ions and organic molecules into the root environment; synthesis of cytokinins and other hormones; penetration of soil layers by mechanical pressure; exploration of new water reservoirs in the soil by root growth. To adequately analyse these root functions it is necessary to isolate the various cell types of the root. Root cell protoplasts can be prepared and fractionated into the various cell types. Protoplast preparation is a standard technique for leaves of many plant species but literature on root cell protoplasts is much scarcer. Fast growing maize roots composed of cells with relatively thin cell walls, present suitable material for isolation of protoplasts (Lin, 1980) and further characterization of ion and sugar transport (Lin *et al.*, 1984). Data on leaf protoplasts should be treated with caution because lipid content and composition may easily be changed by the preparation procedure (Webb & Williams, 1984; Kuiper, 1985). The long preparation procedure required for thick-walled cells, may lead to substantially changed protoplast properties.

The second step, fractionation of the root cells into specific cell types, seems to present even more difficulties. Maize roots may be used for mechanical separation of cortex and stele tissue, and protoplasts may be isolated from both tissues (Lin, 1984). Active ion uptake mainly occurs in the cortex protoplasts.

Clearly the study of root cell types will be long and tedious; but essential for understanding the functioning of intact roots. As an example of cellular control of salt resistance, the role of lipids in the ion exclusion mechanism of rootstocks of 3 Citrus species will be discussed (Douglas & Walker, 1984). Vanadate-sensitive ATPase activity of the plasma membrane preparation of root cells was determined in three species having different ion (Na^+, Cl^-) exclusion capacity, grown at three salinity levels. The best excluder had the highest ATPase activity in the plasma membrane of the root cells. It was further characterized by a low activation energy for enzyme activity and a low phase transition temperature. Salt treatment increased both parameters. An inverse relationship between the activation energy of the plasma membrane ATPase activity and the phospholipid to sterol ratio of the plasma membrane was evident irrespective of genotype and salt treatment (Fig. 1). Clearly, changes in fluidity of the membrane, as induced by free sterols, modulate vanadate-sensitive ATPase activity of the plasma membrane of Citrus root cells; evidently, free sterols are important in the salt exclusion mechanism of Citrus species. The next step for research will be to verify this conclusion by studying preparations of various root cell types.

At this stage it may be wise to warn against a possible misconception that all functions of plant roots are controlled at the cellular level and that consequently all functions may be manipulated by biotechnological experimentation in cell cultures. As will be shown in the following sections, numerous adaptive properties of individual plants are realized at higher organizational levels than that of the cell.

Intact plant roots

The uptake of ions and further translocation to the shoot varies considerably with environmental conditions and, moreover, many species differences are evident. As an example, many halophytes are Na^+ includers and many glycophytes are Na^+ excluders (or K^+ includers). At very low NaCl concentrations *Plantago maritima* absorbs and translocates Na^+ to the shoot (De Boer, 1985). In contrast, the glycophyte *Plantago media* prevents, to a very large extent, Na^+ uptake and further transport to the shoot (Fig. 2). The strategy towards salinity is such that the glycophyte pumps Na^+ ions from the cortex cells to the root environment. In addition, Na^+ is readsorbed from the xylem cells and transported onwards via the root parenchymna cells to the external root environment. The halophyte preferentially absorbs Na^+ and translocates it to the xylem; the roots function in this case as a "K^+ trap" (Fig. 3). Many plant species follow the above strategies for halophytes and glycophytes, but many variations exist. For example, *Spergularia marina*, is a Na^+ includer at low salinity and a Na^+ excluder at high salinity (Cheeseman, 1984).

It is evident that the mechanism of translocation from the root cortical cells to the xylem presents an important goal for future research in salt tolerance and mineral nutrition of plants. In several plant species, electrogenic xylem pumps have been observed (De Boer, 1985), indicating that in plant roots several ion transport ATPases may exist both at the plasmalemma and tonoplast of cortical cells and at the plasmalemma/xylem vessel

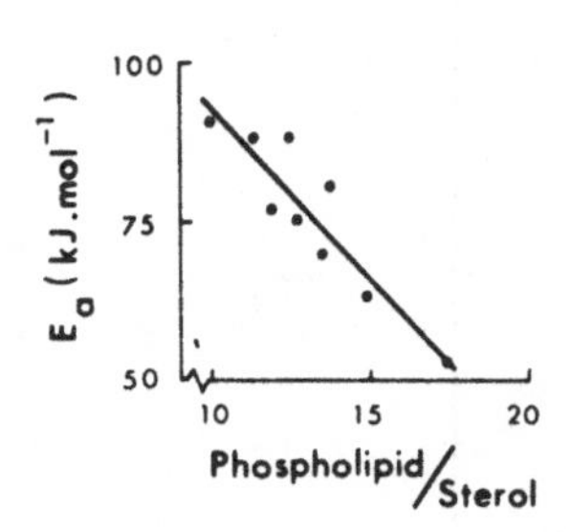

Fig. 1. Relation between the temperature dependence (activation energy) of the plasma membrane ATPase activity of Citrus root cells and the phospholipid to sterol ratio of the same plasma membranes.
The experimental points represent data obtained from 3 cultivars, each grown at 3 salinity levels (Douglas & Walker, 1984).

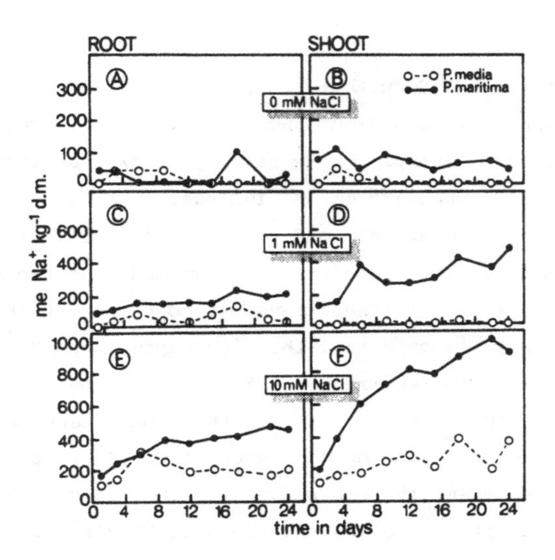

Fig. 2. Time course of Na^+ uptake by the roots and further translocation to the shoot of the salt-sensitive *Plantago media* and the halophyte *Plantago maritima*, studied at 3 salinity levels (De Boer, 1985).

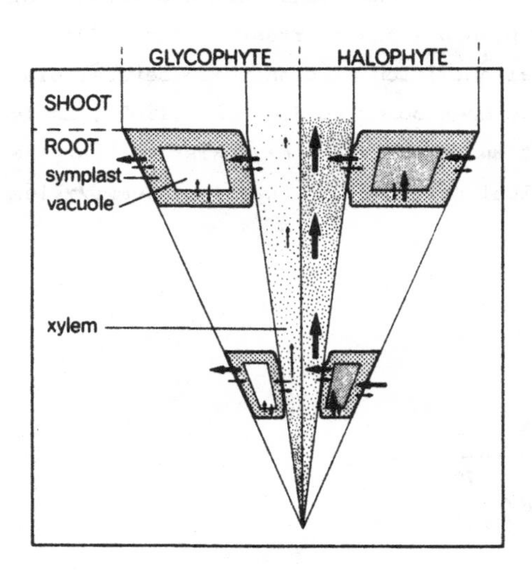

Fig. 3. Model, outlining differences in strategies towards salinity of the glycophyte *Plantago media* and the halophyte *Plantago maritima* (De Boer, 1985).

interface. Fractionation of root cells is required for further analysis of the kinetic properties of these three ATPases.

The involvement of two, spatially separated, electrogenic pumps in plant roots was evident in the study of De Boer, Prins & Zanstra (1983). After excision of the shoot, the following parameters were continuously measured in intact roots: trans-root potential (TRP); xylem water flow, and ion activities (H^+, K^+, Na^+, Cl^-) in the exuding xylem sap. The time course of the TRP of individual plant root systems differed greatly between the root systems studied, and two types of roots, of high (type A) and low TRP (type B), could be distinguished (Fig. 4). Oxygen depletion of the root environment resulted in depolarization of type B roots (low TRP), indicating that O_2 deficiency inhibited a single electrogenic ion pump. The pump is located at the plasmalemma of the epidermal/cortical cells (Fig. 5), and is reactivated again when O_2 supply is restored.

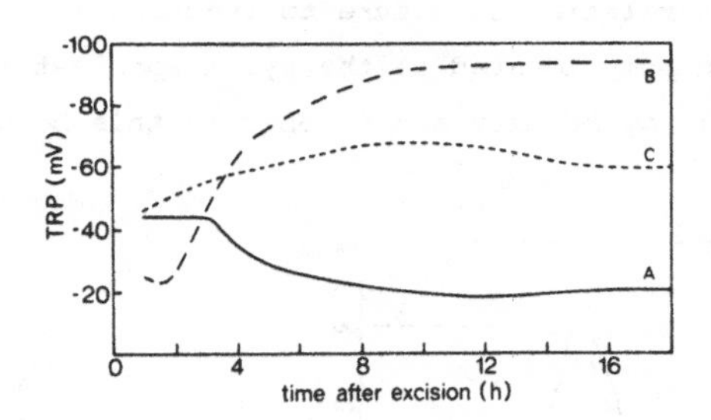

Fig. 4. Time course of trans-root potentials (TRP) in 3 individual root systems, after excision of the shoot. A, type A roots, B, type B roots, C, type C roots (De Boer, Prins & Zanstra, 1983).

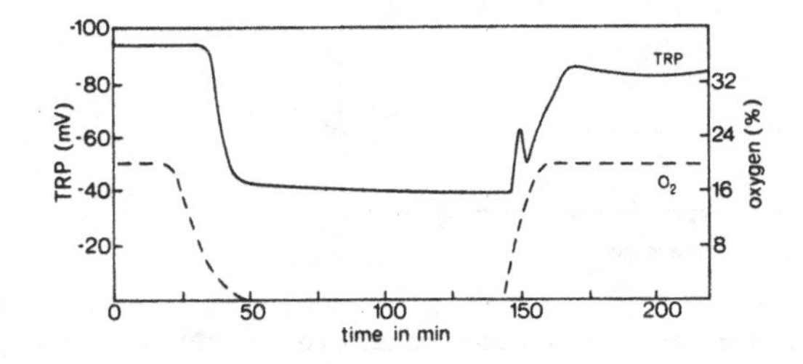

Fig. 5. Reaction of trans-root potential (TRP) by type B roots (low TRP) to changes in oxygen availability of the bathing solution (De Boer *et al.*, 1983).

The reaction of type A roots (high TRP) to anoxia was more complex: a biphasic reaction occurred, so that an initial rapid hyperpolarization was followed by a depolarization (Fig. 6). This result may be explained by the existence of two electrogenic pumps in type A roots, which operate in opposite direction, pumping protons to the outside root environment and to the xylem vessels, respectively. This conclusion is further substantiated by the observation that a decreased O_2 concentration (10%) in the root environment (Fig. 6) inhibited the inner cortical cells/xylem pump; the TRP reacted then as in type B roots.

Two facts are important for future research on ion translocation by such experiments. First, the problem of judging, whether the inner electrogenic pump is functioning (type A) or inactive (type B) in plants grown in water culture. Partial functioning occasionally occurred (Fig. 4, type C). The more heterogenous root systems of plants grown in soil may be possibly more useful for studies relating structure to function as far as the operation of an electrogenic pump located at the xylem apoplast is concerned. The chapters (3 and 5) by McCully and Klepper in this book on

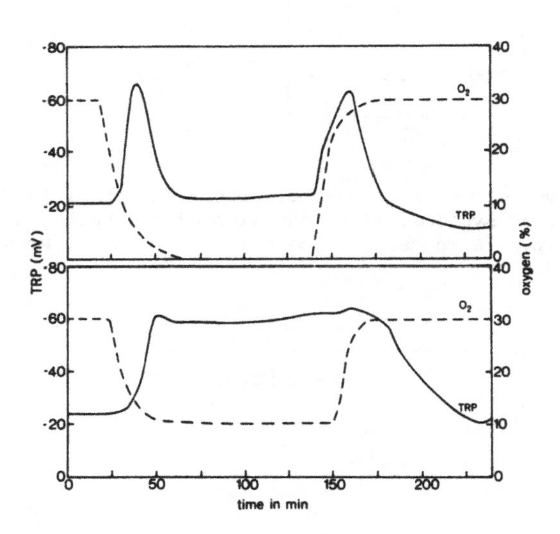

Fig. 6. Reaction of trans-root potential (TRP) by type A roots (high TRP) when roots are brought under anoxia (upper figure) or to limited oxygen (lower figure); De Boer *et al.*, 1983).

the occurrence and branching of lateral roots and the distribution of root xylems in soil should be followed by electrophysiological analyses of the type described above.

Second, both the morphology and the function of roots in ion uptake and translocation to the xylem apoplast are hormonally regulated. In experiments in which *Plantago* roots were perfused by solution containing hormones, it was shown that indoleacetic acid (De Boer *et al.*, 1985), abscisic acid (De Boer, 1985) and fusicoccin (De Boer *et al.*, 1985; De Boer & Prins, 1985) activated the electrogenic pump at the interface between cortical cells and xylem vessels, pumping protons into the xylem and (re)-absorbing K^+ from the xylem into the cortex. Abscisic acid strongly activated xylem sap flow. Even at this early stage of knowledge one may speculate whether an increased level of abscisic acid in the roots, caused by depletion of soil water, may facilitate accumulation of K^+ in the root cortical cells. In this way K^+ would stimulate root growth into hitherto unexploited parts of the soil system. The contributions by McCully and Klepper in this book provide morphological evidence for stimulated lateral root growth under drying conditions; that of Drew (Chapter 4) discussed numerous aspects of root function, water and ion transport, and root morphology.

The intact plant

Mutual interactions between roots and shoots are numerous. One of these interactions, the utilization of carbon in the various functions of roots, is described by Lambers (Chapter 6) in this book. Other more direct effects of the shoot on roots are also known. De Boer (1985) demonstrated that the Na^+ efflux to the xylem, which is so characteristic of halophyte roots, strongly depended on the intactness of the plant; after excision of the shoot it quickly disappeared, indicating the existence of a hormonal signal from the shoot, which induced Na^+ efflux to the root xylem vessels. Also, in wheat roots, the shoot strongly stimulated Na translocation at high salinity (Davis, 1984).

Phytochrome may be directly involved in ion transport. In barley leaves, transport of nitrate from the storage pool to the metabolic pool is mediated by phytochrome (Aslam, Oaks & Hoffaker, 1976). At the same time, nitrate reductase activity is induced by phytochrome. Consequently these processes will affect NO_3^- uptake by the roots and further transport to the shoot.

Another important interaction between roots and shoots is the synthesis of cytokinins in the roots, their translocation to the shoot and their effect on protein synthesis and growth of the shoot. A small nitrate supply to the roots may lead to a low nitrate level in the roots and thus interfere in synthesis of cytokinins (Torrey, 1976). Stress conditions like salinity, drought and flooding reduce cytokinin production in roots (Livne & Vaadia, 1972) and thus modulate shoot growth. Mutual interactions between shoot and roots via cytokinin synthesis (roots) and phytochrome (shoot) are shown schematically in Fig. 7, which includes several factors affecting the phenotypic and adaptive responses involved (Kuiper, 1984a).

The above-mentioned phenotypic responses of individual plants to changes in either the light regime or mineral nutrition (N-supply) or both may vary between species and between genotypes. Genetypic differentiation for phenotypic variation as a response to changes in mineral nutrition has been extensively studied by Kuiper (1984a) in four inbred lines of *Plantago major*. Phenotypic variation for differences in mineral nutrition was determined for growth, photosynthesis, respiration and many enzymes involved in ion transport and N-metabolism. Phenotypic variation was almost absent in the slow-growing line 1 (ssp *major*) and increased growth rate of lines 2, 3 and 4, (ssp *pleiosperma*). Fig. 8 shows the phenotypic variation in shoot:root ratio as a response to changes in the mineral nutrition of the four lines (Kuiper 1984b). Clearly such genotypes provide excellent opportunities to evaluate the role of cytokinins in phenotypic variation in a quantitative way. Immunological techniques for cytokinins are sufficiently advanced at present to test a large number of individual plants.

Ecological aspects

I will comment briefly on the ecological aspects of functioning of plant roots in the physical environment. Ecological research may start with a quantitative evaluation of the plant individuals in a plant community. An evaluation model such as shown in Fig. 7, may be verified for usefulness under field conditions. Such an approach may be attractive because the studied species or genotypes may exhibit different strategies for completion of their life cycle when growing in communities. A good example is shown by *Plantago major* (Kuiper, 1984a). Sub-species *major* (Fig. 8, line 1) has a perennial life cycle, shows hardly any phenotypic responses (except during the seedling stage) exhibits a "K" strategy (Grime, 1977), is competitive, has a low growth rate and must attain a critical size before reproduction. The ssp *pleiosperma* (Fig. 8, line 4) has an annual life cycle, a high degree of

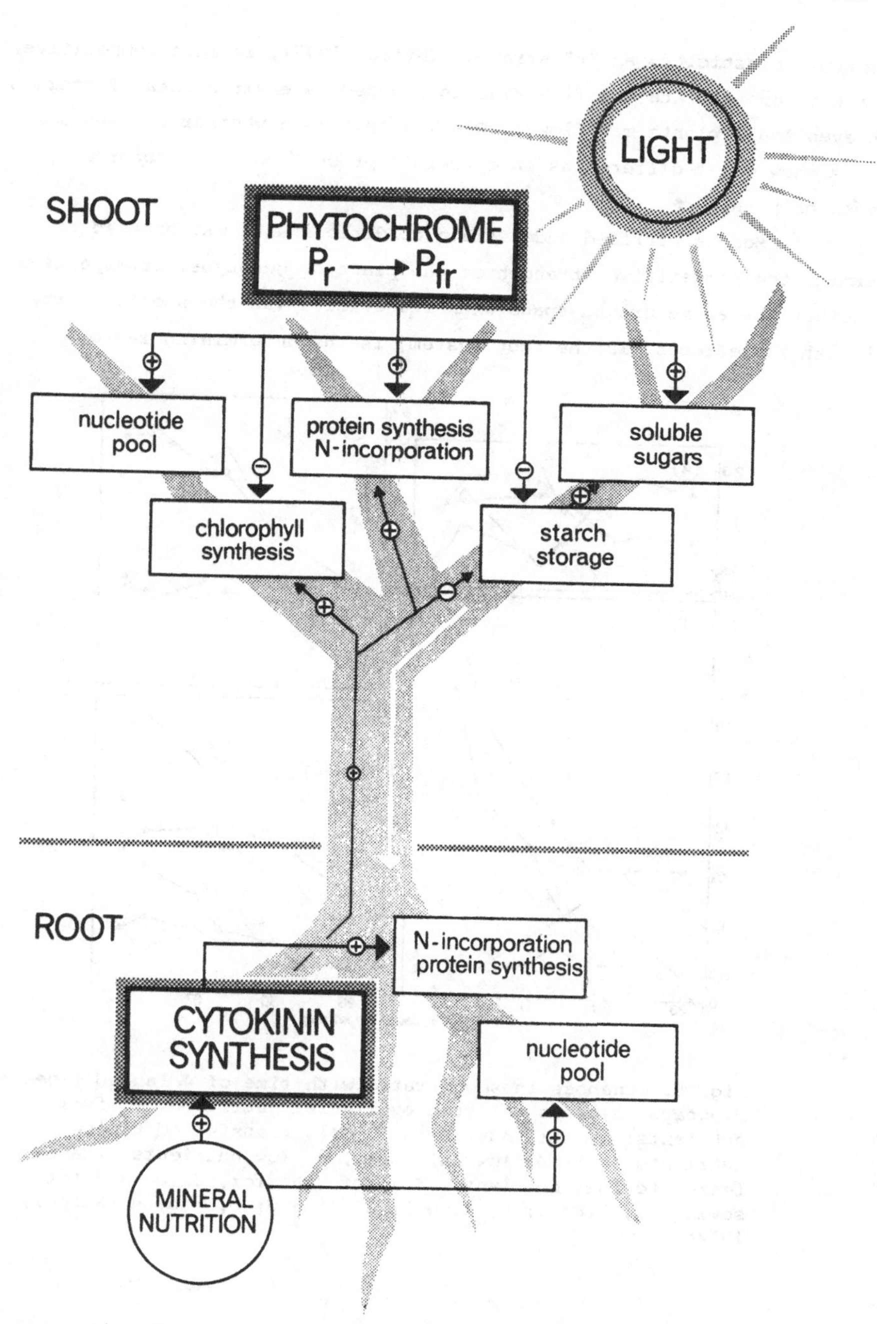

Fig. 7. Model for phenotypic plasticity in higher plants. Mineral nutrition and light are given as stimulating factors (Kuiper, 1984a).

phenotypic plasticity, an "r" strategy (Grime, 1977), is less competitive, has a high growth rate and flowering is induced by environmental factors sc that even small plants may flower. One may speculate whether the two subspecies show great differences in cytokinin production or phytochrome effects or both.

Models verified under field conditions also may be used to determine the competitive strength of the plant or genotype, irrespective of whether the above-ground space with its effects via the shoot, or the soil with its effects via the root system, is the determining factor.

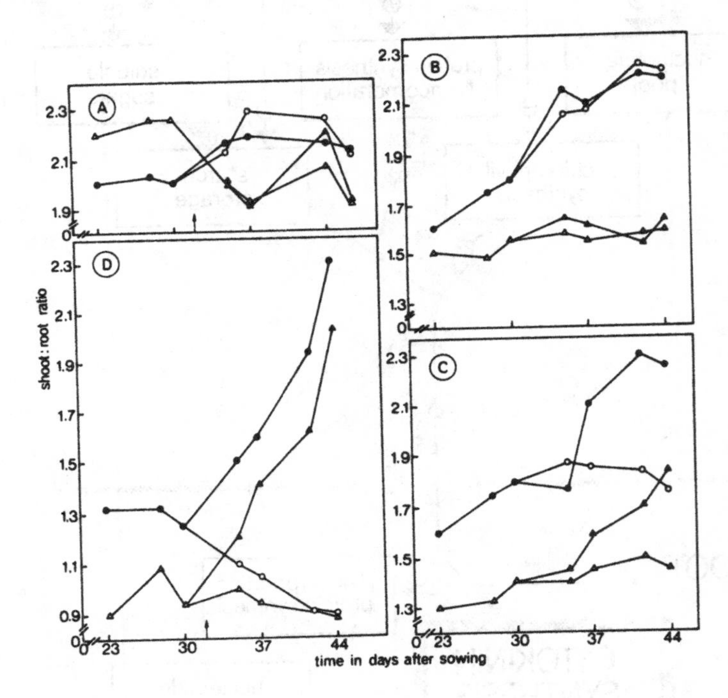

Fig. 8. Changes in shoot:ratio with time of 4 inbred lines of *Plantago major* as affected by mineral nutrition. ●, full nutrients; o, full nutrients (100%), transferred to low nutrients (2%); Δ, low nutrients; ▲, low nutrients, transferred to full nutrients. Time of transfer, 32 days after sowing. A, line 1; B, line 2; C, line 3; D, line 4 (Kuiper, 1984b).

ACKNOWLEDGEMENTS

I thank Drs B. de Boer and D. Kuiper for permission to cite material from their doctoral dissertations at the University of Groningen.

REFERENCES

Aslam, M., Oaks, A. & Huffaker, R.C. (1976). Effect of light and glucose on the induction of nitrate reductase and on the distribution of nitrate in etiolated barley leaves. *Plant Physiology*, 58, 588-91.

Cheeseman, J.M. (1984). Sodium and potassium transport in *Spergularia marina*, a coastal halophyte. *Plant Physiology*, 75, 1023.

Davis, R.F. (1984). A comparison of methods for measuring sodium fluxes in wheat roots. *Plant Physiology*, 75, 1022.

De Boer, A.H. (1985). Xylem/Symplast Ion Exchange: Mechanism and Function in Salt-Tolerance and Growth. Netherlands: Ph.D. Dissertation, University of Groningen.

De Boer, A.H., Prins, H.B.A. & Zanstra, P.E. (1983). Bi-phasic composition of trans-root electrical potentials in roots of *Plantago* species: involvement of spatially separated electrogenic pumps. *Planta*, 157, 259-66.

De Boer, A.H. & Prins, H.B.A. (1985). Xylem perfusion of tap root segments of *Plantago maritima*: the physiological significance of electrogenic xylem pumps. *Plant, Cell and Environment*, 8, 587-94.

De Boer, A.H., Katou, K., Mizuno, A., Kojima, H. & Okamoto, H. (1985). The role of electrogenic xylem pumps in K^+ absorption from the xylem of *Vigna unguiculata*: the effects of auxin and fusicoccin. *Plant, Cell and Environment*, 8, 579-86.

Douglas, T.J. & Walker, R.R. (1984). Phospholipids, free sterols and adenosine triphosphatase of plasma membrane-enriched preparations from roots of *Citrus* genotypes differing in chloride exclusion ability. *Physiologia Plantarum*, 62, 51-8.

Grime, J.P. (1977). Evidence for the existence of three primary strategies in plants and its relevance to ecological and evolutionary theory. *American Naturalist*, 111, 1169-94.

Kuiper, D. (1984a). Genetic Differentiation and Phenotypic Plasticity in *Plantago* Species. Netherlands: Ph.D. Dissertation, University of Groningen.

Kuiper, D. (1984b). Genetic differentiation in *Plantago major*: growth and root respiration and their role in phenotypic adaptation. *Physiologia Plantarum*, 57, 222-30.

Kuiper, P.J.C. (1985). Environmental changes and lipid metabolism of higher plants. *Physiologia Plantarum*, 64, 118-22.

Lin, W. (1980). Corn root protoplasts. Isolation and general characterization of ion transport. *Plant Physiology*, 66, 550-4.

Lin, W. (1984). Comparison of ion transport properties of protoplasts isolated from cortex and stele tissue for corn roots. *Plant Physiology*, 75, 1025.

Lin, W., Schmitt, M.R., Hitz, W.D. & Giaquinta, R.T. (1984). Sugar transport in isolated corn root protoplasts. *Plant Physiology*, 76, 894-8.

Livne, A. & Vaadia, Y. (1972). Water deficits and hormone relations. In *Water-deficits and plant growth*, vol. 3, ed. T.T. Kozlowski, p pp. 255-76. New York: Academic Press.

Torrey, J.G. (1976). Root hormones and plant growth. *Annual Review of Plant Physiology*, 27, 435-59.
Webb, M.S. & Williams, J.P. (1984). Changes in the lipid and fatty acid composition of *Vicia faba* mesophyll protoplasts induced by isolation.*Plant and Cell Physiology*, 25, 1541-50.

INDEX

SOCIETY FOR EXPERIMENTAL BIOLOGY SEMINAR SERIES

A series of multi-author volumes developed from seminars held by the Society for Experimental Biology. Each volume serves not only as an introductory review of a specific topic, but also introduces the reader to experimental evidence to support the theories and principles discussed, and points the way to new research.

1. Effects of air pollution on plants Edited by T.A. MANSFIELD
2. Effects of pollutants on aquatic organisms Edited by A.P.M. LOCKWOOD
3. Analytical and quantitative methods Edited by J.A. MEEK and H.Y. ELDER
4. Isolation of plant growth substances Edited by J.R. HILLMAN
5. Aspects of animal movement Edited by H.Y. ELDER and E.R. TRUEMAN
6. Neurones without impulses: their significance for vertebrate and invertebrate systems Edited by A. ROBERTS and B.M.H. BUSH
7. Development and specialisation of skeletal muscle Edited by D.F. GOLDSPINK
8. Stomatal physiology Edited by P.G. JARVIS and T.A. MANSFIELD
9. Brain mechanisms of behaviour in lower vertebrates Edited by P.R. LAMING
10. The cell cycle Edited by P.C.L. JOHN
11. Effects of disease on the physiology of the growing plant Edited by P.G. AYRES
12. Biology of the chemotactic response Edited by J.M. LACKIE and P.C. WILLIAMSON
13. Animal migration Edited by D.J. AIDLEY
14. Biological timekeeping Edited by J. BRADY
15. The nucleolus Edited by E.G. JORDAN and C.A. CULLIS
16. Gills Edited by D.F. HOULIHAN, J.C. RANKIN and T.J. SHUTTLEWORTH
17. Cellular acclimatisation to environmental change Edited by A.R. COSSINS and P. SHETERLINE
18. Plant biotechnology Edited by S.H. MANTELL and H. SMITH
19. Storage carbohydrates in vascular plants Edited by D.H. LEWIS
20. The physiology and biochemistry of plant respiration Edited by J.M. PALMER
21. Chloroplast biogenesis Edited by R.J. ELLIS
22. Instrumentation for environmental physiology Edited by B. MARSHALL and F.I. WOODWARD
23. The biosynthesis and metabolism of plant hormones Edited by A. CROZIER and J.R. HILLMAN
24. Coordination of motor behaviour Edited by B.M.H. BUSH and F. CLARAC
25. Cell ageing and cell death Edited by I. DAVIES and D.C. SIGEE
26. The cell division cycle in plants Edited by J.A. BRYANT and D. FRANCIS
27. Control of leaf growth Edited by N.R. BAKER, W.J. DAVIES and C. ONG
28. Biochemistry of plant cell walls Edited by C.T. BRETT and J.R. HILLMAN
29. Immunology in plant science Edited by T.L. WANG
30. Root development and function Edited by P.J. GREGORY, J.V. LAKE & D.A. ROSE
31. Plant canopies: their growth, form and function Edited by G. Russell, B. MARSHALL & P.G. JARVIS (forthcoming)
32. Developmental mutants in higher plants Edited by H. THOMAS & D. GRIERSON